Derivative-Free Adaptive Control

Derivative-Free Adaptive Control

Tansel Yucelen, Associate Professor

University of South Florida, Department of Mechanical Engineering, Tampa, FL

Anthony J. Calise, Professor Emeritus

Georgia Institute of Technology, School of Aerospace Engineering, Atlanta, GA

Volume 267
Progress in Astronautics and Aeronautics

Timothy C. Lieuwen, Editor-in-Chief
Georgia Institute of Technology
Atlanta, Georgia

AIAA
American Institute of Aeronautics and Astronautics, Inc.

Cover photo: AirStar flight test vehicle. Courtesy: NASA Langley Research Center.

ISBN 978-1-62410-701-6
Ebook ISBN 978-1-62410-702-3

PROGRESS IN ASTRONAUTICS AND AERONAUTICS

TABLE OF CONTENTS

PREFACE

Our purpose in preparing this monograph has been threefold. The first is to collect in a single document all of the fundamental aspects of what we have written previously regarding the development of and theory for adaptive control, which we refer to as derivative-free model reference adaptive control (DF-MRAC). These developments began when the first author was a PhD student in the School of Aerospace Engineering with the second author as his advisor. Our first joint publication on this subject was in the form of an AIAA conference paper in the Guidance, Navigation, and Control Conference in August 2011. All previous methods of model reference adaptive control (MRAC) were based on the underlying assumption that plant model uncertainty can be approximated by a set of state-dependent basis functions, which are parameterized by a set of unknown but constant weights. The stability analysis that emerges results in a differential equation, which is integrated over time to provide an estimate for the so-called ideal values for the weights. Furthermore, if the system is not sufficiently excited, then there is no guarantee the estimated weights will converge to the ideal weights. DF-MRAC takes the viewpoint that ideal weights themselves may be time varying, and in fact (as we shall show), one can even do away with the need for choosing a set of basis functions.

The second purpose is to bring the reader's attention to shortcomings in MRAC, which are often ignored in publications but are critically important when it comes to certification and transitioning this technology into fielded systems such as aircraft and guided munitions. These shortcomings include upset recovery due to sudden failures which are difficult to parameterize, robustness to time delay and unmodeled dynamics; the effect sensor noise has on adaptive control; and the conflict that arises when attempting to adaptively augment a nominal controller having proportional plus integral action.

The third purpose of this monograph is to present a set of reproducible examples illustrating the important features and advantages of DF-MRAC over conventional MRAC methods, and to point out the many circumstances for which a constant parameterization of uncertainty may be difficult or impossible to achieve, particularly in the case of unmodeled dynamics, time-varying uncertainties, and exogenous disturbances that may be acting on the system being controlled.

ACKNOWLEDGEMENTS

We wish to acknowledge the financial support provided by the Air Force Office of Scientific Research (AFOSR), Air Force Research Laboratory (AFRL), and National Aeronautics and Space Administration (NASA) during the course of our research efforts, and for the opportunities provided by Georgia Tech in performing laboratory and flight experiments and by NASA in flight testing our research on its AirStar aircraft.

DEDICATION

To our parents.

CHAPTER 1

Introduction

1.1 BACKGROUND

Every dynamic system, without exception, is subject to exogenous disturbances and/or parametric and dynamic uncertainties. In aircraft, for example, exogenous disturbances arise from winds and turbulences. The presence of parametric and/or dynamic uncertainties occurs when idealized assumptions, such as linearization or model order reduction, are introduced or when degraded modes of operation occur due to either actuator failure or structural damage. In general, disturbances and uncertainties are regarded as anomalies tending to degrade stability and performance of a physical system, particularly when operated under feedback control.

When designing a control law for a dynamic system, a fundamental problem is to maintain a desired level of stability and performance in the presence of anomalies. Similar in spirit to robust control methods, adaptive control methods rely on theoretical tools to address this problem. In contrast to robust control methods, however, these methods have the capability to correct for, and in some cases nearly cancel, the effect of a specified class of anomalies in real time, through adaptation mimicking a learning process. This implies nominal performance need not necessarily be compromised in return for robustness to the degrading effects of anomalies. Moreover, in some cases adaptive control methods can even improve system performance in response to exogenous disturbances over time.

Adaptive control methods are classified as either direct or indirect, and can be further subdivided into state feedback or output feedback. In this monograph we focus on model reference adaptive control (MRAC) architectures, a well-known class of direct adaptive control methods. Among the excellent textbooks on MRAC methods, we refer the reader in particular to [1–9] and the relatively recent survey articles in [10–13]. In MRAC, the desired closed-loop system response is represented in the form of a reference model whose state (or output) is compared with the state (or output) of the uncertain dynamical system. The error signal resulting from this comparison drives the parameter adjustment mechanism operating in real-time for the purpose of regulating the uncertain dynamical system to follow the behavior of the reference model. Depending on the nature of the anomalies, the behavioral error can be shown to either asymptotically vanish or remain bounded with an adjustable bound.

MRAC approaches deal with multivariable uncertain systems in continuous-time. These approaches are derived based on Lyapunov stability theory and either assume or derive a weight update law in the form of an ordinary differential equation for the weight estimates. All these methods have in common the underlying assumption there exists an unknown, but constant, ideal set of weights that parameterize uncertainty. Although this assumption seems reasonable, and many MRAC architectures may work well, in some applications they can require the use of unrealistically high adaptation gain, or fail to achieve a desired level of performance in terms of recovery from damages and/or actuator failure. MRAC laws requiring high gain can excite unmodeled dynamics, typically exhibit an excessive amount of control activity [11, 12], amplify the effect of sensor noise, and are not sufficiently robust to time delay [13].

1.2 PURPOSE AND CONTENT

We refer to all the adaptive methods in [1–13] as "derivative-based." One purpose in this monograph is to explain in detail an alternative approach to adaptation and learning and identify significant advantages over the approaches in these references and others; we refer to this alternative approach as derivative-free MRAC (DF-MRAC). We provide the theoretical foundation and develop a family of approaches for DF-MRAC that can be applied to a variety of problems, including both state feedback and output feedback forms, and illustrate its advantages both in the setting of simple problem formations and in a variety of more realistic problems in flexible satellite attitude control, aircraft flight control, and control of large-scale flexible systems. These advantages are largely afforded by the so-called "ideal weights," which in DF-MRAC are permitted to be time varying, and because a time delay used in implementation adds an important additional dimension to the tuning process.

Our development of DF-MRAC has been motivated from the start by the fact that, in spite of the many years of development, there are very few instances in which adaptive control has been transitioned into a fielded system. We believe the reason lies partly within the issues we have raised in this monograph, such as the effect of sensor noise, time delay, robustness to external disturbances, unwanted interactions when augmenting a nominal controller with integral action, and the inability to respond properly to uncertainties due to unexpected failures. In aircraft applications, handling qualities require measures of robustness, like gain and phase margins, that are well defined for fixed gain controller design. Such accepted measures are lacking in adaptive control. To our knowledge, the only full transition to a fielded system is a design implemented on the Joint Direct Attack Munition (JDAM) guided munition [14]. The highly gained scheduled JDAM autopilot has since been replaced with a fully adaptive autopilot. This has resulted in a dramatic reduction in the cost and development time whenever the military wanted to apply the JDAM tail kit to a different bomb. This required design and retuning of a three dimensional table of gains involving a year of tunnel testing and repeated drop testing at Eglin AFB. The need for gain tuning has been completely eliminated now that the autopilot has been made fully adaptive.

Chapter 2 provides a brief summary of MRAC in the context of a simple problem formulation in state feedback form, introduces some of the notation employed in the remainder of the monograph, and concludes with several simple examples illustrating some of the weaknesses of MRAC. Chapter 3 begins by introducing DF-MRAC in state feedback form, provides a theoretical justification for derivative-free adaptation, considers several means by which modification terms may be introduced, and shows how the method extends to the case of exogenous disturbances. It also introduces a specialized form applicable to both MRAC and DF-MRAC relying on the notion of adaptive loop-transfer recovery (ALR), especially useful when applied to systems subjected to anomalies including both unmodeled dynamics and time delay. It concludes by presenting a series of simple examples highlighting the various features of DF-MRAC. Chapter 4 addresses the important problem of unmodeled dynamics. It shows how DF-MRAC has a natural robustness to unmodeled dynamics not present in MRAC. Chapter 5 addresses the problem of DF-MRAC in output feedback form. Chapter 6 discusses the application of DF-MRAC to large-scale, interconnected, uncertain systems. It shows how it is particularly well suited for disturbance rejection, as well as for adaptation in the presence of sudden change in each subsystem's uncertain dynamics, such as might be caused by damage and/or when subsystems are interconnected. Chapter 7 is devoted to illustrating several practical applications in flight control and includes some flight test results. A summary of all the major conclusions we have drawn from our research into DF-MRAC to date is given in Chapter 8, and several directions in which this research may be extended are outlined.

1.3 NOTATION

R (R_+)	The set of real (positive) values
R^n	The set of $n \times 1$ real column vectors
$R^{n\times m}$	The set of $n \times m$ real matrices
$(\blacksquare)^{\mathrm{T}}$	Transpose
$(\blacksquare)^{-1}$	Inverse
$\triangleq$	Equality by definition
I_n	Identity matrix of dimension $n \times n$
0_n	n-element row or column vector of zeros
$\lambda(M)$	Eigenvalues of M

$\underline{\lambda}(M)$, $(\overline{\lambda}(M))$	Minimum (maximum) eigenvalue of M
$\underline{\sigma}(M)$, $(\overline{\sigma}(M))$	Minimum (maximum) singular value of M
$\lvert\cdot\rvert$	Euclidian vector norm
$\lVert\cdot\rVert$	Frobenius matrix norm
vec(■)	Column stacking operator
$tr[A]$	Trace of matrix A
diag$[\lambda(A)]$	Diagonal matrix formed with the eigenvalues of A
diag$[A,B]$	Block diagonal matrix formed with matrices A and B
D_+^{nxm} $(\overline{D}_+^{nxm})$	The set of $n \times n$ diagonal positive-definite (semidefinite) real matrices with diagonal scalar entries

1.4 ABBREVIATIONS

ALR	adaptive loop-transfer recovery
DF-MRAC	derivative-free model reference adaptive control
GPS	global positioning system
GTM	generic transport model
LQG	linear quadratic Gaussian
LTR	loop transfer recovery
MRAC	model reference adaptive control
P+I	proportional plus integral
SPR	strictly positive real
UUB	uniformly ultimately bounded

REFERENCES

[1] Kokotovic, P. V., *Foundations of Adaptive Control*, Springer, Berlin, Heidelberg, 1991.

[2] Krstic, M., Kanellakopoulos, I., and Kokotovic, P. V., *Nonlinear and Adaptive Control Design*, Wiley, 1995.

[3] Tao, G., *Adaptive Control Design and Analysis (Vol. 37)*, John Wiley & Sons, 2003.

[4] Ioannou, P., and Fidan, B., *Adaptive Control Tutorial*, Society for Industrial and Applied Mathematics, 2006.

[5] Hovakimyan, N., and Cao, C., *L1 Adaptive Control Theory: Guaranteed Robustness with Fast Adaptation*, SIAM, 2010.

[6] Sastry, S., and Bodson, M., *Adaptive Control: Stability, Convergence and Robustness*, Courier Corporation, 2011.

[7] Narendra, K. S., and Annaswamy, A. M., *Stable Adaptive Systems*, Courier Corporation, 2012.

[8] Astrom, K. J., and Wittenmark, B., *Adaptive Control*, Courier Corporation, 2013.

[9] Lavretsky, E., and Wise, K. A., *Robust and Adaptive Control with Aerospace Applications*, Springer, 2013.

[10] Wise, K. A., Lavretsky, E., and Hovakimyan, N., "Adaptive Control of Flight: Theory, Applications, and Open Problems," American Control Conference, IEEE, 2006, pp. 5966–5971.

[11] Anderson, B. D., and Dehghani, A., "Challenges of Adaptive Control—Past, Permanent and Future," *Annual Reviews in Control*, Vol. 32, No. 2, 2008, pp. 123–135.

[12] Black, W. S., Haghi, P., and Ariyur, K. B., "Adaptive Systems: History, Techniques, Problems, and Perspectives," *Systems*, Vol. 2, No. 4, 2014, pp. 606–660.

[13] Tao, G., "Multivariable Adaptive Control: A Survey," *Automatica*, Vol. 50, No. 11, 2014, pp. 2737–2764.

[14] Calise, A. J., Sharma, M., and Corban, J. E. "Adaptive Autopilot Design for Guided Munitions," *AIAA Journal of Guidance, Control, and Dynamics*, Vol. 23, No. 5, 2000.

CHAPTER 2

Model Reference Adaptive Control

2.1 A SIMPLIFIED PROBLEM FORMULATION

In this section, we state standard results for the derivative-based MRAC problem in state feedback form. Consider the uncertain system given by

$$\dot{x}(t) = Ax(t) + B[u(t) + \Delta(x(t))] \tag{2.1}$$

where $x(t) \in R^n$ is the state vector, $u(t) \in R^m$ is the control input, $A \in R^{nxn}$ and $B \in R^{nxm}$ are known matrices, and $\Delta : R^n \rightarrow R^m$ is a matched uncertainty. We assume the pair (A, B) is controllable, $x(t)$ is available for feedback, and $u(t)$ is restricted to the class of admissible controls consisting of measurable functions.

The reference model is given by

$$\dot{x}_m(t) = A_m x_m(t) + B_m r(t) \tag{2.2}$$

where $x_m(t) \in R^n$ is the reference model state vector, $A_m \in R^{nxn}$ is Hurwitz, $B_m \in R^{nxr}$, and $r(t) \in R^r$ with $r \leq m$ is a bounded piecewise continuous reference input. Because $r(t)$ is bounded, it follows that $x_m(t)$ is uniformly bounded for all $x_m(0)$. This defines the desired behavior of the system operating in closed loop for a set of prescribed reference inputs.

Assumption 2.1: The matched uncertainty in Eq. (2.1) can be linearly parameterized as

$$\Delta(x) = W^T \beta(x) + \varepsilon(x), \quad |\varepsilon(x)| \leq \varepsilon^*, \quad x \in D_x \tag{2.3}$$

where $W \in R^{sxm}$ is the unknown *constant* weight matrix, $\beta(x) : R^n \rightarrow R^s$ is a known vector of basis functions of the form $\beta(x) = [\beta_1(x), \beta_2(x), \ldots, \beta_s(x)]^T$, $\varepsilon(x) : R^n \rightarrow R^m$ is the residual error, and D_x is a sufficiently large compact set $D_x \subset R^n$.

2.2 DERIVATIVE-BASED ADAPTATION

Consider the following feedback control law:

$$u(t) = u_n(t) - u_{ad}(t) \tag{2.4}$$

where $u_n(t)$ is a nominal feedback control given by

$$u_n(t) = -K_1 x(t) + K_2 r(t) \tag{2.5}$$

where $K_1 \in R^{mxn}$ and $K_2 \in R^{mxr}$ are nominal control gains such that $A - BK_1$ is Hurwitz and $u_{ad}(t)$ is the adaptive feedback control component given by

$$u_{ad}(t) = \hat{W}^T(t)\beta(x(t)) \tag{2.6}$$

where $\hat{W}(t) \in R^{sxm}$ is an estimate of W satisfying the derivative-based weight update law

$$d\hat{W}/dt = \gamma\left[\beta(x(t))e^T(t)PB + dW_m/dt\right], \quad \gamma > 0 \tag{2.7}$$

with the error

$$e(t) \triangleq x(t) - x_m(t) \tag{2.8}$$

defined as the state tracking error. In Eq. (2.7), the matrix $P \in R^{nxn}$ is the positive-definite solution of the Lyapunov equation

$$0 = A_m^T P + PA_m + Q \tag{2.9}$$

for any given $Q = Q^T > 0$, and $dW_m/dt \in R^{sxm}$ is a modification term, for example,

$$dW_m/dt = -\sigma\hat{W}(t) \tag{2.10}$$

in the case of σ-modification [1] or

$$dW_m/dt = -\sigma|e(t)|\hat{W}(t) \tag{2.11}$$

in the case of e-modification [2], where $\sigma > 0$ is the modification gain.

Assumption 2.2: A_m and B_m in Eq. (2.2) are chosen so that

$$A_m = A - BK_1 \tag{2.12}$$

$$B_m = BK_2 \tag{2.13}$$

Using Eqs. (2.3–2.6), (2.12), and (2.13), the dynamics in Eq. (2.1) can be written as

$$\dot{x}(t) = A_m x(t) + B_m r(t) + B\,\tilde{W}^T(t)\beta(x(t)) + B\varepsilon(x(t)) \tag{2.14}$$

where

$$\tilde{W}(t) \triangleq W - \hat{W}(t) \tag{2.15}$$

is the weight update error. The state tracking error can likewise be written as

$$\dot{e}(t) = A_m e(t) + B\tilde{W}^T(t)\beta(x(t)) + B\varepsilon(x(t)) \tag{2.16}$$

Remark 2.1: As previously noted, it is essential K_1 be chosen so that $A_m = A - BK_1$ is stable. In addition, it is desirable that a regulated output $y(t) \in R^r$, defined by $y(t) = Cx(t)$, follows $r(t)$. To ensure this occurs, K_2 is chosen to satisfy $C(A - BK_1)^{-1}BK_2 = -I_r$.

2.3 STABILITY ANALYSIS

Theorems providing sufficient conditions under which the closed-loop system errors $e(t)$ and $\tilde{W}(t)$ are uniformly ultimately bounded (UUB) for the σ- and e-modification cases can be found in [1, 2]. In what follows, we consider the more simplified case in which modification terms are not present and employ the Lyapunov function candidate

$$L(e(t), \tilde{W}(t)) = e(t)^T Pe(t) + \gamma^{-1} tr\left\{\tilde{W}^T(t)\tilde{W}(t)\right\} \tag{2.17}$$

Further assume the parameterization of uncertainty in Eq. (2.3) is exact, or $\varepsilon^* = 0$. The time derivative of Eq. (2.17) is

$$\dot{L}(t) = e(t)^T P\dot{e}(t) + \dot{e}(t)Pe(t) + 2tr\left\{\tilde{W}^T(t)\dot{\tilde{W}}(t)\right\} \tag{2.18}$$

Using Eqs. (2.7), (2.9), and (2.16), this becomes

$$\dot{L}(t) = -e(t)^T Qe(t) + 2e^T PB\tilde{W}^T\beta - 2tr\left\{\tilde{W}^T(t)\beta e^T PB\right\}$$

Because $tr\left\{\tilde{W}^T(t)\beta e^T PB\right\} = e^T PB\tilde{W}^T\beta$, it follows that

$$\dot{L}(t) = -e(t)^T Qe(t) \leq 0 \tag{2.19}$$

This ensures the pair $(e(t), \tilde{W}(t))$ is bounded for all $t > 0$. Therefore, $\dot{e}(t)$ in Eq. (2.16) is also bounded, which from Eq. (2.19) implies $\ddot{L}(t)$ is bounded. It therefore follows from Barbalat's lemma [3] that

$$\lim_{t\to\infty} \dot{L}(t) = 0 \tag{2.20}$$

and consequently Eq. (2.19) further implies

$$\lim_{t\to\infty} e(t) = 0 \tag{2.21}$$

Note $\lim_{t\to\infty} \tilde{W}(t) = 0$ is guaranteed in this case only if the conditions for so-called "persistence of excitation" are satisfied [3]. This is a direct consequence of the fact $\dot{L}(t)$ in Eq. (2.19) is only negative semidefinite. Parameter convergence is normally not of concern in the design of an adaptive control system unless there is also a requirement for online system identification, in which case it depends largely on the nature of $r(t)$.

Remark 2.2: The preceding stability analysis can easily be extended to the case of matched input anomalies by modifying Eq. (2.1) to the form $\dot{x}(t) = Ax(t) + B\Lambda[u(t) + \Delta(x(t))]$, in which $\Lambda > 0$ is an unknown diagonal matrix. The treatment of this extension is covered in Section 3.4.

2.4 TWO EXAMPLES OF DEFICIENCIES IN DERIVATIVE-BASED MRAC

In this section we illustrate two deficiencies in derivative-based MRAC using a simple first-order example of aircraft wing-rock [4]. The first example illustrates what can happen in the case of time-varying ideal weights; the second example is one in which the ideal weights are constant but the nominal controller contains integral action. These examples will later be revisited in the context of DF-MRAC in Chapter 3.

For the first example, the open loop system dynamics are given by

$$\dot{x}(t) = L_p x(t) + L_\delta[u(t) + \Delta(x(t))] \tag{2.22}$$

where $x(t)$ represents roll rate, $u(t)$ represents aileron deflection, $L_p = -1$ is the roll damping coefficient, $L_\delta = 1$ is the aileron control effectiveness, and $\Delta(x(t)) = w(t)x(t)$, with $w(t)$ being a square wave having an amplitude of 1.0 and a period of 5.0 seconds. Therefore, it is constant when viewed piecewise in time. Although uncertainty in aircraft dynamics does not normally behave in this manner, aircraft stability derivatives can undergo a sudden change in the event of damage to the airframe. So this example should be regarded in this context. The reference model parameters are: $A_m = -2$ and $B_m = 2$. These choices define nominal performance by ensuring a zero steady state error in response to a reference step command, and a 5% settling time of 1.5 seconds in the absence of anomalies. The corresponding nominal feedback gains in Eqs. (2.12) and (2.13) are $K_1 = 1$ and $K_2 = 2$.

Figure 2.1 shows the performance of the nominal control design in the presence of uncertainty for the case in which the reference input, $r(t)$ in Eq. (2.5), is also a square wave. In Fig. 2.2, the derivative-based MRAC control law in Eq. (2.7) (without modification) with $\beta(x(t)) = x(t)$ is used with adaptation gains $\gamma = 10^2$ and $\gamma = 10^4$, respectively, and with P determined using $Q = 1$ in Eq. (2.9). The lower value for γ was tuned to provide the best reasonable response. Tracking performance is improved by increasing the adaptation gain beyond 10^2; however, this causes high-frequency oscillations in the control response, which would be unacceptable in a real system.

In the second example the ideal weight is constant, and the uncertainty is chosen as $\Delta(x(t)) = 1 + 5x(t)$ in Eq. (2.2), which now includes a bias term. To

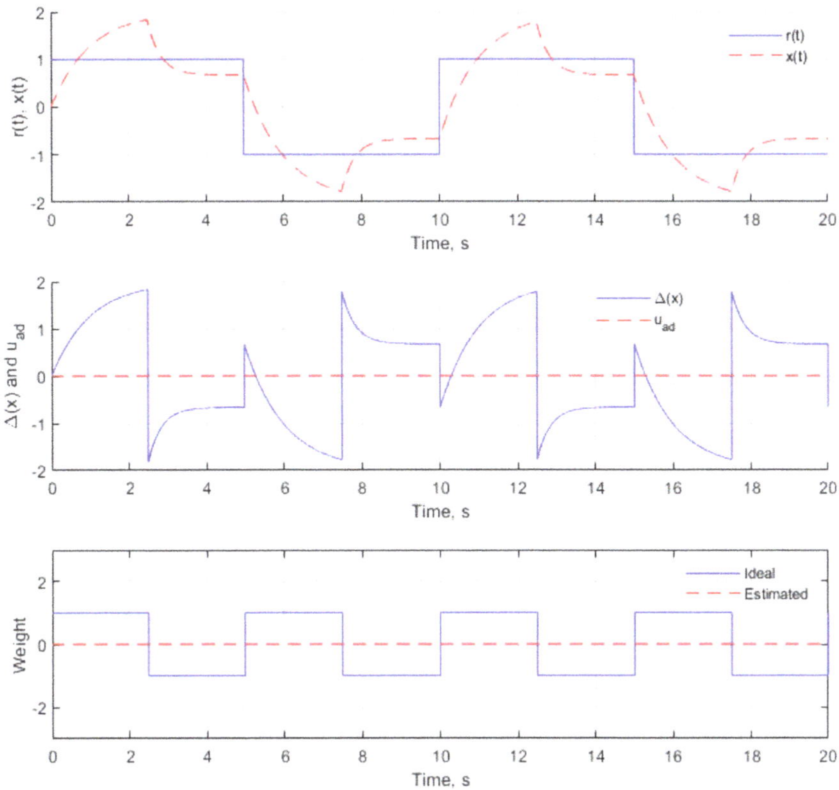

Fig. 2.1 Nominal controller responses for an uncertainty with square wave ideal weight.

overcome the effect of this bias, the nominal control design includes an integral term. The uncertainty bias is purposely included to illustrate the point that conventional MRAC is problematic when augmenting a nominal controller having integral action. We therefore introduce an integrator state

$$\dot{x}_i(t) = -x(t) + r(t) \tag{2.23}$$

The dynamics become

$$\dot{x}_a(t) = \begin{bmatrix} L_p & 0 \\ -1 & 0 \end{bmatrix} x_a(t) + \begin{bmatrix} L_\delta \\ 0 \end{bmatrix} (u(t) + \Delta(x)) + \begin{bmatrix} 0 \\ 1 \end{bmatrix} r(t) \tag{2.24}$$

where $x_a^T(t) = [x(t)\ x_i(t)]$ is the augmented state vector. The proportional plus integral (P+I) form for the nominal controller is

$$u_n(t) = -[k_x \quad k_i] x_a(t) \tag{2.25}$$

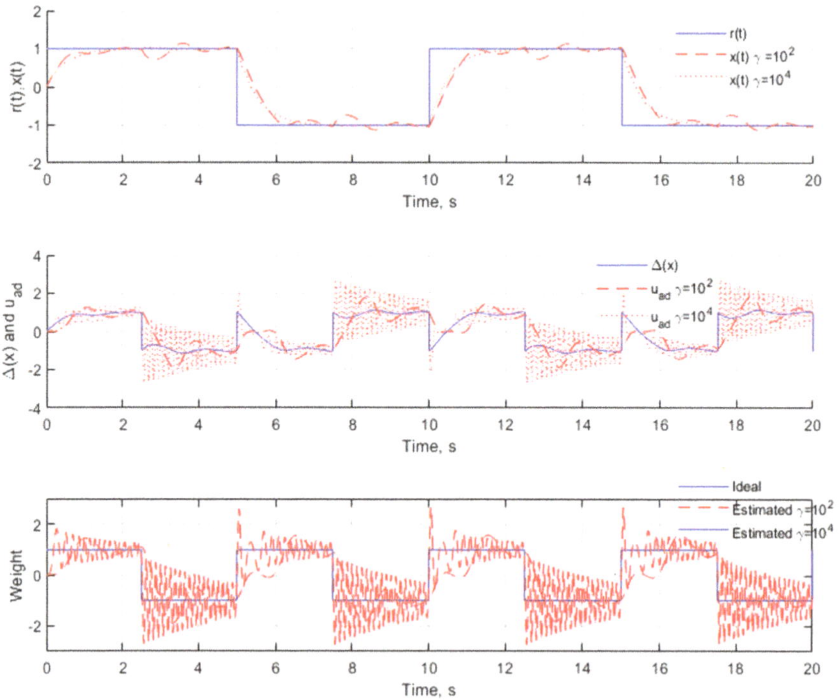

Fig. 2.2 MRAC responses using $\gamma = 10^2$ and $\gamma = 10^4$.

and the corresponding reference model in Eq. (2.2) becomes

$$\dot{x}_m(t) = \begin{bmatrix} L_p - L_\delta K_x & -L_\delta K_i \\ -1 & 0 \end{bmatrix} x_m + \begin{bmatrix} 0 \\ 1 \end{bmatrix} r(t) \tag{2.26}$$

The P+I gains are set to $K_x = 1.5$ and $K_i = -3$, which corresponds to a nominal system having a damping ratio of 0.72 and a natural frequency of 1.73 rad/s. In this case, the nominal controller is unstable in the presence of uncertainty (not shown). The adaptive control is chosen in the form of Eq. (2.6), with $\beta^T(x(t)) = \begin{bmatrix} 1 & x_a^T(t) \end{bmatrix}$, which now includes a bias term. We again apply the adaptive law in Eq. (2.7) (without the modification term) and with P determined using $Q = I_2$ in Eq. (2.9), but this time with adaptation gains chosen as $\gamma = 20$ and $\gamma = 200$, where the lower value is tuned to provide the best achievable response. Figure 2.3 illustrates the conflict that can arise when a nominal control law with integral action is augmented with an MRAC controller employing integration in its weight update law. Note the slowly varying tracking error during the first transient phase. In more complex problems, this slow variation can be persistent. Introducing σ-modification makes matters increasingly worse. Figure 2.4 shows what happens when the same MRAC controller is employed

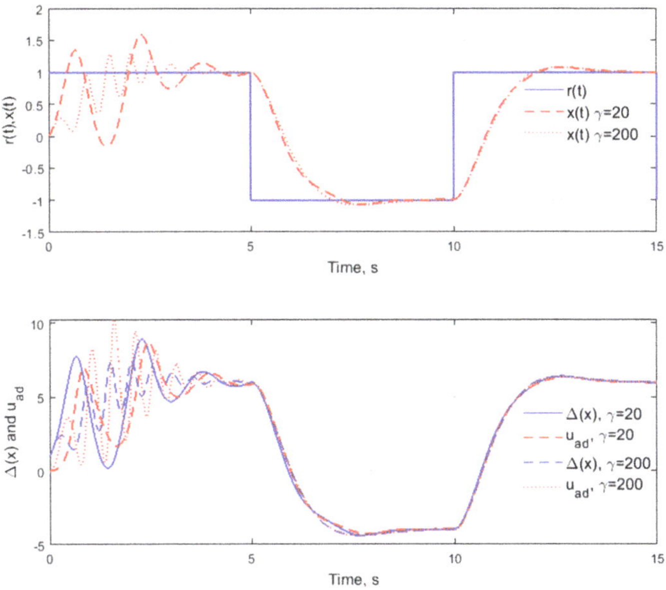

Fig. 2.3 MRAC responses for a controller with integral action and an uncertainty with constant ideal weight.

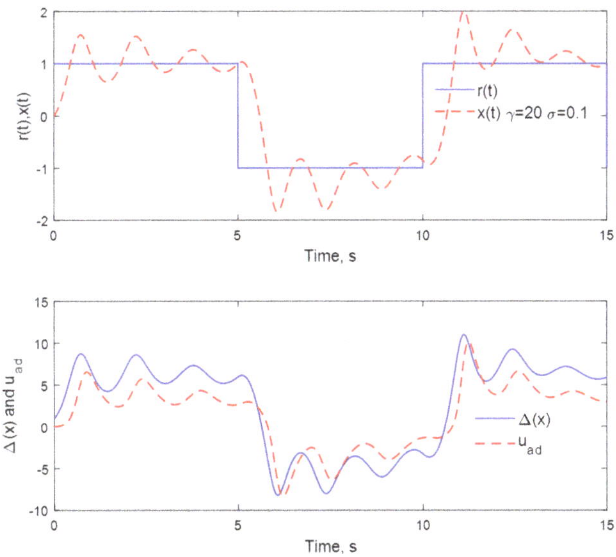

Fig. 2.4 MRAC with σ-modification responses for a controller with integral action and an uncertainty with constant ideal weight.

with σ-modification using $\sigma = 0.1$ for the $\gamma = 20$ case in Fig. 2.3. Further increases in σ only worsen the response. A similar behavior is obtained with e-modification. A direct comparison with DF-MRAC for this case appears in Chapter 3 (Fig. 3.9).

REFERENCES

[1] Ioannou, P., and Kokotovic, P., "Instability Analysis and Improvement of Robustness of Adaptive Control," *Automatica*, Vol. 20, No. 5, 1984, pp. 583–594.

[2] Narendra, K., and Annaswamy, A., "A New Adaptive Law for Robust Adaptation Without Persistent Excitation," *IEEE Transactions on Automatic Control*, Vol. 32, No. 2, 1987, pp. 134–145.

[3] Hassan, K., *Nonlinear Systems*, 3rd ed., Prentice Hall, Upper Saddle River, NJ, 2002.

[4] Hsu, C.-H., and Lan, C. E., "Theory of Wing Rock," *Journal of Aircraft*, Vol. 22, No. 10, 1985, pp. 920–929.

CHAPTER 3

Derivative-Free Model Reference Adaptive Control

3.1 MOTIVATION

In this chapter we develop a derivative-free model reference adaptive control (DF-MRAC) law [1] that uses the information of delayed weight estimates and the information contained in current system states and state errors. The proposed method is an extension of the iterative learning approach adopted in [2] for purposes of adaptive observer design. We relax the assumption of constant unknown ideal weights to the existence of time-varying weights. Fast and possibly discontinuous variation in weights is allowed. We prove the error dynamics are uniformly ultimately bounded (UUB) using a Lyapunov–Krasovskii functional. We consider constant unknown ideal weights as a special case and prove the state tracking error dynamics in this case are asymptotically stable. Finally, we discuss employing various modification terms for further improving the performance and robustness of the adaptively controlled system.

DF-MRAC differs from MRAC because it does not make use of integration in its weight update law. In fact, DF-MRAC challenges the conventional wisdom of expecting an adaptive law to cancel the effect of matched constant disturbances and uncertainties, because one does not need adaptation to correct for biases. Biases can more reliably be handled using a nonadaptive controller with integral action. This ensures errors in the regulated output variables go to zero for constant disturbances and constant model errors even if they are unmatched, so long as the closed loop system remains stable. Therefore, because many existing guidance and flight controllers do employ integral action, and if the goal is to augment an existing controller, then the weight update law should not have the effect of integral action; otherwise, a conflict may arise resulting in a slowly varying tracking error (as previously illustrated in Figs. 2.3 and 2.4). In these circumstances, the role of adaptation should be confined to things like maintaining stability, error transient performance, fast upset recovery in the event there is a destabilizing failure, and preserving to the extent possible the time delay margins of the nominal design. DF-MRAC is particularly well suited for these circumstances.

This chapter will highlight the application of DF-MRAC to systems that can undergo a sudden change in dynamics, such as might be due to reconfiguration, deployment of a payload, docking, or structural damage. We will also address the problem of robustness to time delay. Other advantages, such as robustness to unmodeled dynamics will be covered in Chapter 4, and reduction of sensitivity to sensor noise, will be discussed in Chapter 5.

3.2 THE STATE FEEDBACK CASE

The following assumption strengthens Assumption 2.1 in Chapter 2 by setting $\varepsilon(x(t)) = 0$, which can be justified because time variation is allowed in the unknown weight matrix.

Assumption 3.1: The matched uncertainty in Eq. (2.1) can be linearly parameterized as

$$\Delta(x(t)) = W^T(t)\beta(x(t)), \quad x \in D_x \tag{3.1}$$

where $W(t) \in R^{sxm}$ is a *time-varying* weight matrix satisfying $\|W(t)\| \leq w^*$, and $\beta(x(t)) : R^n \rightarrow R^s$ is a vector of basis functions of the form $\beta(x) = [\beta_1(x), \beta_2(x), \ldots, \beta_s(x)]^T \in R^s$.

Remark 3.1: Assumption 3.1 expands the class of uncertainties that can be represented by any given set of basis functions. An adaptive law designed subject to Assumption 3.1 can be more effective than an adaptive law designed subject to Assumption 2.1 in canceling a wider class of uncertainties, due to the fact time variation is allowed in the unknown weight matrix.

Remark 3.2: Assumption 3.1 does not place any restriction on the time derivative of the weight matrix; however, the degree of time dependence will depend on how $\beta(x)$ is chosen.

The following theorem presents the main result of this chapter.

Theorem 3.1: Consider the uncertain system given by Eq. (2.1) subject to Assumption 3.1. Consider, in addition, the feedback control law given by Eq. (2.4), with the nominal feedback control component given by Eq. (2.5) subject to Assumption 2.2, and with the adaptive feedback control component given by Eq. (2.6) having a derivative-free weight update law in the form

$$\hat{W}(t) = \Omega_1 \hat{W}(t-\tau) + \hat{\Omega}_2(t), \; t > \tau \tag{3.2}$$

where $\tau > 0$, $\Omega_1 \in R^{sxs}$, and $\hat{\Omega}_2(t) \; : \; R^n x R^n \; \rightarrow \; R^{sxm}$ satisfy

$$0 \leq \Omega_1^T \Omega_1 < I_s \tag{3.3}$$

$$\hat{\Omega}_2(t) = \; \kappa_2 \beta(x(t)) e^T(t) PB, \kappa_2 > 0 \tag{3.4}$$

with $P \in R^{nxn}$ satisfying Eq. (2.9) for any symmetric $Q > 0$. Then, $e(t)$ and $\tilde{W}(t)$ are UUB.

Proof: Using Eq. (3.2) and defining

$$\Omega_2(t) \triangleq W(t) - \Omega_1 W(t-\tau) \tag{3.5}$$

where $\|\Omega_2(t)\| \leq \delta^*$, $\delta^* \triangleq w^*(1+\|\Omega_1\|)$, the weight update error in Eq. (2.15), with W replaced by $W(t)$, can be rewritten as

$$\tilde{W}(t) = \Omega_1 \tilde{W}(t-\tau) + \Omega_2(t) - \hat{\Omega}_2(t) \tag{3.6}$$

Using Eq. (3.6), the state tracking error dynamics in Eq. (2.16) under Assumption 3.1 becomes

$$\dot{e}(t) = A_m e(t) + B\left[\Omega_1 \tilde{W}(t-\tau) - \hat{\Omega}_2(t) + \Omega_2(t)\right]^T \beta(x(t)) \tag{3.7}$$

To show the closed-loop system given by Eqs. (3.6) and (3.7) is UUB, consider the Lyapunov–Krasovskii functional [3]

$$L_k(e(t),\ \tilde{W}_t) = e^T(t)Pe(t) + \rho\, tr\left[\int_{t-\tau}^{t} \tilde{W}^T(s)\tilde{W}(s)ds\right],\quad \rho = 1/\kappa_2 \tag{3.8}$$

where $\tilde{W}_t$ represents $\tilde{W}(t)$ over the time interval $t-\tau$ to t. The directional derivative of Eq. (3.8) along the closed-loop system trajectories of Eqs. (3.6) and (3.7) is given by

$$\begin{aligned}\dot{L}_k(e(t),\tilde{W}_t) = &-e^T(t)Qe(t) + 2e^T(t)PB\left[\Omega_1\tilde{W}(t-\tau)\right]^T\beta(x(t))\\ &-2e^T PB\hat{\Omega}_2^T(t)\beta(x(t)) + 2e^T(t)PB\Omega_2^T(t)\beta(x(t))\\ &+\rho tr\left[\tilde{W}^T(t)\tilde{W}(t) - \tilde{W}^T(t-\tau)\tilde{W}(t-\tau)\right]\end{aligned} \tag{3.9}$$

Using Eq. (3.6) to expand the term $tr[\tilde{W}^T(t)\ \tilde{W}(t)]$ in Eq. (3.9) produces

$$\begin{aligned}\dot{L}_k(e(t),\tilde{W}_t) = &-e^T(t)Qe(t) + 2e^T(t)PB\left[\Omega_1\tilde{W}(t-\tau)\right]^T\beta(x(t))\\ &-2e^T(t)PB\hat{\Omega}_2^T(t)\beta(x(t)) + 2e^T(t)PB\Omega_2^T(t)\beta(x(t))\\ &+\rho tr[\tilde{W}^T(t-\tau)\Omega_1^T\Omega_1\tilde{W}(t-\tau) + \hat{\Omega}_2^T(t)\hat{\Omega}_2(t)\\ &+\Omega_2^T(t)\Omega_2(t) - 2\hat{\Omega}_2^T(t)\Omega_1\tilde{W}(t-\tau) + 2\tilde{W}^T(t-\tau)\Omega_1^T\Omega_2(t)\\ &-2\hat{\Omega}_2^T(t)\Omega_2(t) - \tilde{W}^T(t-\tau)\tilde{W}(t-\tau)]\end{aligned} \tag{3.10}$$

Next, consider the fact $a^T b \leq \gamma a^T a + b^T b/4\gamma,\ \gamma > 0$, which follows from Young's inequality [4] for vectors a and b. This can be generalized to matrices as $tr\left[A^T B\right] = \text{vec}(A)^T\text{vec}(B) \leq \gamma\text{vec}(A)^T\text{vec}(A) + \text{vec}(B)^T\text{vec}(B)/4\gamma = \gamma\,\text{tr}\left[A^T A\right] + \text{tr}\left[B^T B\right]/4\gamma,\ \gamma >$, for any matrices A

and B having appropriate dimensions. Using this, we can write

$$tr\left[2\tilde{W}^T(t-\tau)\Omega_1^T\Omega_2(t)\right] \le tr\left[\gamma\tilde{W}^T(t-\tau)\Omega_1^T\Omega_1\tilde{W}(t-\tau)\right] + tr\left[\Omega_2^T(t)\Omega_2(t)/\gamma\right], \quad \gamma > 0 \tag{3.11}$$

Using Eq. (3.4) for $\hat{\Omega}_2(t)$, imposing $\rho = 1/\kappa_2$, and substituting the right-hand side of Eq. (3.11) for the ninth term in Eq. (3.10), it follows that

$$\begin{aligned}\dot{L}_k(e(t),\tilde{W}_t) \le -e^T(t)Qe(t) - \rho tr\left[\tilde{W}^T(t-\tau)\left\{I_s - (1+\gamma)\Omega_1^T\Omega_1\right\}\tilde{W}(t-\tau)\right] \\ + \rho(1+1/\gamma)tr\left[\Omega_2^T(t)\Omega_2(t)\right]\end{aligned} \tag{3.12}$$

In arriving at Eq. (3.12), the condition $\rho = 1/\kappa_2$ in Eq. (3.8) and the adaptation law in Eq. (3.4) combine to ensure the second and eighth terms in Eq. (3.10) cancel. The same is true for the 4th and 10th terms. Using Eq. (3.4), the third term can be equivalently expressed as $-2\rho tr\left[\hat{\Omega}_2^T(t)\hat{\Omega}_2(t)\right]$; therefore, the third and sixth terms combine to produce a term that is at least negative semidefinite. The second term in Eq. (3.12) is a combination of the fifth and last terms in Eq. (3.10) with the first term in Eq. (3.11). The last term in Eq. (3.12) results from combining the seventh term in Eq. (3.10) with the second term in Eq. (3.11). This allows us to reduce Eq. (3.12) to the form

$$\dot{L}_k(e(t),\tilde{W}_t) \le -c_1|e(t)|^2 - c_2\|\tilde{W}^T(t)\|^2 - c_3\|\tilde{W}^T(t-\tau)\|^2 + d \tag{3.13}$$

where

$$c_1 = \underline{\lambda}(Q) > 0 \tag{3.14}$$

$$c_2 = 0 \tag{3.15}$$

$$c_3 = \rho\underline{\lambda}(I_s - (1+\gamma)\Omega_1^T\Omega_1) > 0, \quad 0 < \gamma < \bar{\gamma} \tag{3.16}$$

$$d = \rho(1+1/\gamma)\delta^{*2} > 0 \tag{3.17}$$

and it becomes necessary to restrict γ to lie in the open interval $(0, \bar{\gamma})$ in Eq. (3.16). Note the existence $\bar{\gamma} > 0$ in Eq. (3.16) is assured by Eq. (3.3) in Theorem 3.1, which ensures $c_3 > 0$ when $\gamma = 0$. From Eq. (3.13), it follows that $|e(t)| > \Psi_1$ or $\|\tilde{W}^T(t-\tau)\| > \Psi_3$ renders $\dot{L}_k(e(t),\tilde{W}_t) < 0$, where

$$\Psi_1 = \delta^*\sqrt{(1+1/\gamma)/\underline{\lambda}(Q)\kappa_2} > 0 \tag{3.18}$$

$$\Psi_3 = \delta^*\sqrt{(1+1/\gamma)/\underline{\lambda}(I_s - (1+\gamma)\Omega_1^T\Omega_1)} > 0, \quad 0 < \gamma < \bar{\gamma} \tag{3.19}$$

Hence $e(t)$ and $\tilde{W}(t-\tau)$ are UUB. This is true for all t; therefore, $\tilde{W}_t(t)$ is UUB. ■

The adaptive control architecture is shown in Fig. 3.1.

Remark 3.3: The forms in Eqs. (3.18) and (3.19) suggest $\Psi_1(\kappa_2)$ can be made arbitrarily small by choosing an arbitrarily large value for κ_2, while having no effect on Ψ_3; however, as shown in the

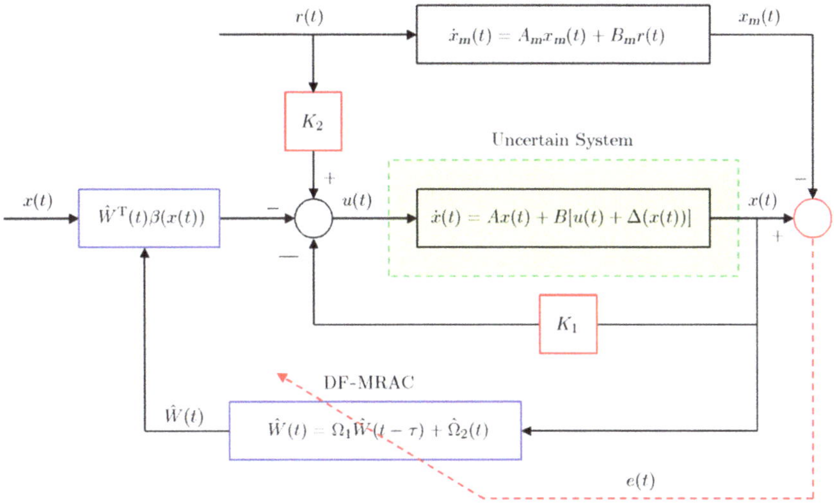

Fig. 3.1 Visualization of the DF-MRAC architecture.

following corollary, using a large value for κ_2 has no effect on the ultimate bound. Also note $\Psi_3(\gamma)$ reaches a minimum for some value of γ within the open interval $(0, \bar{\gamma})$ because it is infinite at both boundaries for γ.

Define $q(t) \triangleq [e^T(t), \tilde{v}(t)]^T$, where $\tilde{v}^2(t) \triangleq tr\left[\int_{t-\tau}^{t} \tilde{W}^T(s)\tilde{W}(s)ds\right]$, and let $B_r = \{q(t) : |q(t)| < r\}$ such that $B_r \subset D_q$ for a sufficiently large compact set D_q. Then, we have the following corollary:

Corollary 3.1: Under the conditions of Theorem 3.1, an estimate for the ultimate bound, for the case $\xi > 0$, is given by

$$r = \sqrt{\frac{\bar{\lambda}(P)\Psi_1^2 + \rho\tau\Psi_3^2}{\underline{\lambda}(\bar{P})}} \tag{3.20}$$

where $\bar{P} = \mathrm{diag}[P, \rho]$.

Proof: Denote $\Omega_\alpha = \{q(t) \in B_r : q^T(t)\bar{P}q(t) \leq \propto\}$, $\propto = \min_{\|q(t)=r\|} q^T(t)\bar{P}q(t) = r^2\underline{\lambda}(\bar{P})$. Because

$$L_k(e(t), \tilde{W}_t) = q^T(t)\bar{P}q(t) = e^T Pe(t) + \rho\, tr\left[\int_{t-\tau}^{t} \tilde{W}^T(s)\tilde{W}(s)ds\right] \tag{3.21}$$

if follows Ω_α is an invariant set as long as

$$\alpha \geq \bar{\lambda}(P)\Psi_1^2 + \rho\tau\Psi_3^2 \tag{3.22}$$

Thus the minimum size of B_r, ensuring this condition has a radius given by Eq. (3.20). ∎

The sets used in this proof are illustrated in Fig. 3.2.

Remark 3.4: Because $\rho = 1/\kappa_2$, we have $\underline{\lambda}(\bar{P}) = \rho$ when $\kappa_2 > 1/\underline{\lambda}(P)$. It also can be seen that Ψ_1^2 in Eq. (3.18) is also proportional to ρ for all values of κ_2. Therefore, when $\kappa_2 > 1/\underline{\lambda}(P)$, the ultimate bound expression in Eq. (3.20) is independent of κ_2.

Remark 3.5: Letting $dW_m/dt = -\sigma\hat{W}(t)$ in Eq. (2.7) and using a first-order Euler method for integration, with τ_s being the step size, results in

$$\hat{W}(t) = [\tau_s\gamma\beta(x(t))e^T(t)PB + \hat{W}(t-\tau_s)]/(1+\tau_s\gamma\sigma) \tag{3.23}$$

This form of weight update law for MRAC is identical to the DF-MRAC law in Eq. (3.2) if $\Omega_1 = I_s/(1+\tau_s\gamma\sigma)$, $\kappa_2 = \tau_s\gamma/(1+\tau_s\gamma\sigma)$, and $\tau = \tau_s$. In DF-MRAC, Ω_1 can be chosen, for example, as ςI_s where $0 < \varsigma < 1$, but τ is not necessarily equal to τ_s. This added dimension in the tuning process provides greater flexibility and increased memory to the adaptive law.

Remark 3.6: The fact that DF-MRAC and MRAC with sigma modification are related in form by Eq. (3.23) helps to explain why the DF-MRAC weight update law given by Eq. (3.2) subject to Eqs. (3.3) and (3.4) does not explicitly require a modification term to prove the error dynamics, including the weight errors, are UUB even in the presence of unmodeled dynamics (see Chapter 4).

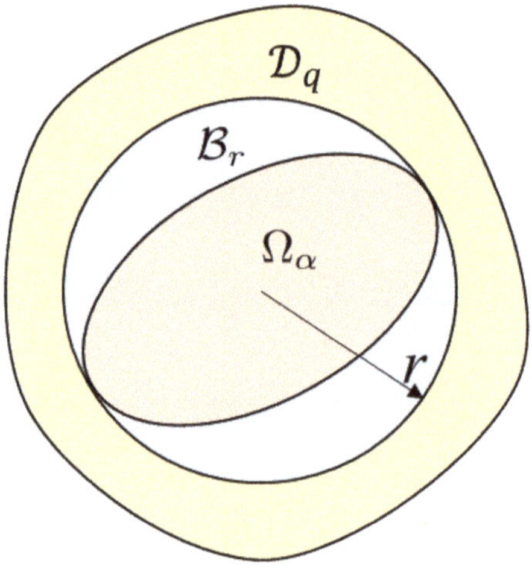

Fig. 3.2 Geometric representation of sets.

Remark 3.7: What takes place when implementing Eq. (3.2) when $t < \tau$ is up to the designer. For example, one might choose $\hat{W}(t) = 0$ or $\hat{W}(t) = \hat{\Omega}_2(t)$. We have assumed the system remains controllable, even for large values of τ. In any practical system, the effect of control saturation comes into play, and the adaptive control may not be able to recover from too large an upset.

For the case of constant ideal weights in Assumption 3.1, Theorem 3.1 specializes to the following theorem. In this case, we assume the uncertainty is structured; that is, $\beta(x(t))$ in Eq. (3.1) represents the vector of *known* basis functions in Eq. (2.3).

Theorem 3.2: Consider the uncertain system given by Eq. (2.1) subject to Assumption 3.1, where $W \in R^{s\times m}$ is an unknown *constant* weight matrix. Consider, in addition, the feedback control law given by Eq. (2.4), with the nominal feedback control component given by Eq. (2.5) subject to Assumption 2.2, and with the adaptive feedback control component given by Eq. (2.6) having the derivative-free weight update law in the form of Eqs. (3.2) and (3.4), with $\Omega_1 = I_s$. Then, $e(t)$ and $\tilde{W}(t)$ approach a subspace in these error variables in which $e(t) = 0$ and $B\tilde{W}^T(t)\beta(x(t)) = 0$.

Proof: The result follows directly from the proof of Theorem 3.1 by choosing $\Omega_1 = I_s$. In this case, $\delta^* = 0$ due to $\Omega_2(t) = 0$ in Eq. (3.5), which follows from the fact the ideal weights are constant, the uncertainty is structured, and $\Omega_1 = I_s$. Then, the inequality in Eq. (3.11) is not needed because the left side vanishes. In this case, $\gamma = 0$, $c_3 = 0$ in Eq. (3.16), and $d = 0$ in Eq. (3.17). Therefore, it follows from Eq. (3.13) that the entire error space is invariant. Let $\mathcal{E}$ denote the set of points in this space where $\dot{L}_k(e(t), \tilde{W}(t)) = 0$. From Eq. (3.13), all points in $\mathcal{E}$ lie in a subspace where $e(t) = 0$, and it follows from LaSalle's theorem (see, e.g., [5]) that all solutions in the error space approach the largest invariant set $\mathcal{M}$ in $\mathcal{E}$. Now $e(t) = 0$ implies $\hat{\Omega}_2(t) = 0$ and $\dot{e}(t) = 0$. Then, Eq. (3.7), with all of the above taken together, implies $\mathcal{M}$ is composed of all points in the error space in which $e(t) = 0$ and $B\tilde{W}^T(t)\beta(x(t)) = 0$. ∎

Remark 3.8: The system is said to be sufficiently excited if the reference input *r(t)* is such that the conditions $e(t) = 0$ and $B\tilde{W}^T(t)\beta(x(t)) = 0$ admit only the solution $\tilde{W}(t) = 0$ in the limit $t \to \infty$. This amounts to the standard MRAC condition for persistency of excitation [6].

Remark 3.9: DF-MRAC does not employ an integrator in its weight update law. This is advantageous from the perspective of augmenting a nominal controller employing integral action to ensure the regulated output variables track the reference $r(t)$ in the presence of constant disturbances, *regardless* of how these disturbances may enter the system. An example illustrating this advantage is provided in Section 3.4.

3.3 INTRODUCING MODIFICATION TERMS

Although the derivative-free weight update law does not require a modification term to prove the error dynamics are UUB, one may wish to employ a modification term in order to improve performance or robustness of the system. The following theorem extends Theorem 3.1 to a more general form of modified DF-MRAC.

Theorem 3.3: Consider the uncertain system given by Eq. (2.1), subject to Assumption 3.1. Consider, in addition, the feedback control law given by Eq. (2.4), with the nominal feedback control component given by Eq. (2.5), subject to Assumption 2.2, and with the adaptive feedback control component given by Eq. (2.6) having a derivative-free weight update law in the form given by Eq. (3.2), where $\tau > 0$ is a time delay design value, $\Omega_1 \in R^{s\times s}$, and $\hat{\Omega}_2(t)$ is replaced with $\hat{\Omega}_{2m}(t)$, satisfying

$$0 \leq \Omega_1^T \Omega_1 < I_s \tag{3.24}$$

$$\hat{\Omega}_{2m}(t) = \kappa_2 \left[\beta(x(t)) e^T PB - \kappa_m S(t) \hat{W}(t) \right], \kappa_m > 0, \quad S(t) \geq 0 \tag{3.25}$$

where κ_m is the modification gain, and $S(t) \in R^{sxs}$. Then, $e(t) = x(t) - x_m(t)$ and $\tilde{W}(t) = W(t) - \hat{W}(t)$ are UUB.

Proof: Replacing $\hat{\Omega}_2(t)$ in Eq. (3.2) with $\hat{\Omega}_{2m}(t)$ and solving for $\hat{W}(t)$ produces

$$\hat{W}(t) = \Omega_{1m} \hat{W}(t - \tau) + \kappa_{2m} \hat{\Omega}_2(t) \tag{3.26}$$

where

$$\Omega_{1m}(t) = [I_s + \kappa_2 \kappa_m S(t)]^{-1} \Omega_1 \tag{3.27}$$

$$\kappa_{2m}(t) = [I_s + \kappa_2 \kappa_m S(t)]^{-1} \kappa_2 \tag{3.28}$$

Then the proof follows directly from the proof of Theorem 3.1 with $\Omega_{1m}(t)$ replacing $\Omega_1(t)$ at every step, except that it remains to be shown the modified condition

$$c_{3m}(t) = \rho \underline{\lambda}(I_s - (1+\gamma)\Omega_{1m}^T(t)\Omega_{1m}(t)) > 0, \quad 0 < \gamma < \bar{\gamma} \tag{3.29}$$

is assured by the condition for c_3 in Eq. (3.16), or equivalently that the condition in Eq. (3.16) ensures Eq. (3.29) is true $\forall t$. We begin by noting the following properties of matrix singular values:

$$\bar{\sigma}(A+B) \leq \bar{\sigma}(A) + \bar{\sigma}(B) \tag{3.30a}$$

$$\bar{\sigma}(AB) \leq \bar{\sigma}(A)\bar{\sigma}(B) \tag{3.30b}$$

$$\bar{\sigma}(A^{-1}) = 1/\bar{\sigma}(A) \tag{3.30c}$$

The first property is the so-called triangle inequality and is common to all matrix norms. Furthermore,

$$\underline{\lambda}(I - A^T A) = \min_{|x|=1} x^T (I - A^T A) x = 1 - \max_{|x|=1} (x^T A^T A x) = 1 - \bar{\sigma}^2(A) \tag{3.31}$$

Therefore the condition in Eq. (3.29) can be written as

$$c_{3m}(t) = \rho(1 - (1+\gamma)\bar{\sigma}^2(\Omega_{1m}(t)) > 0 \tag{3.32}$$

Applying the properties in Eq. (3.30) to the expression for Ω_{1m} in Eq. (3.27), we have

$$\bar{\sigma}(\Omega_{1m}) \leq \bar{\sigma}(\Omega_1)/\bar{\sigma}(I_s + \kappa_2\kappa_m S(t)) \leq \bar{\sigma}(\Omega_1) \tag{3.33}$$

because $S(t) \geq 0$. Therefore, Eq. (3.32) is assured by replacing Ω_{1m} with Ω_1, and by Eq. (3.31) this is equivalent to the condition in Eq. (3.16). ■

Remark 3.10: $S(t) = |e(t)|I_s$ in the case of e-modification [7]. In the case of σ-modification [8], $S(t) = I_s$.

Remark 3.11: An adaptive loop-transfer recovery (ALR) modification term was developed in [9] with the objective of recovering the loop transfer properties of the reference system. ALR modification is applied by choosing $S(t) = \beta_x(x(t))\beta_x^T(x(t))$, where $\beta_x(x(t)) \triangleq d\beta(x(t))/dx(t) \in R^{sxn}$. Examples illustrating the use of ALR modification to improve response to time delay are given in Sections 3.5.4 and 7.2.

3.4 EXTENSION TO THE CASE OF INPUT ANOMALIES

Consider the case of input uncertainty confined to the range space of B by writing the uncertain system in the form

$$\dot{x}(\mathrm{t}) = Ax(\mathrm{t}) + B\Lambda[u(t) + \Delta(x(t))] \tag{3.34}$$

where $\Lambda \in R^{mxm}$ is an unknown diagonal matrix with diagonal elements $\lambda_i > 0,\ i = 1, \ldots, m$, representing uncertainty in control effectiveness. Once again, as in Eq. (2.4), we let

$$u(t) = u_n(t) - u_{ad}(t) \tag{3.35}$$

Using Eq. (3.35), the requirement that $u_{ad}(t)$ ideally cancels the uncertainty in Eq. (3.34) requires

$$\Lambda[u_n(t) - u_{ad}(t) + \Delta(x(t))] = u_n(t) \tag{3.36}$$

Solving for $u_{ad}(t)$ in Eq. (3.36) produces the condition

$$u_{ad}(t) = \Delta(x(t)) + \Lambda^{-1}[\Lambda - I_m]u_n(t) = \Delta(x(t)) + Du_n(t) \tag{3.37}$$

where $D \in R^{mxm}$ is also diagonal. Therefore, the extension of all the previous results in this chapter to the case of input anomalies follows by defining

$$u_{ad}(t) = \hat{W}^T(t)\beta(x(t)) + \hat{D}u_n(t) \tag{3.38}$$

The diagonal elements of $\hat{D}$ satisfy

$$\hat{d}_i(t) = \varphi_{d_i}\hat{d}_i(t - \tau_{d_i}) + \hat{\omega}_{d_i}(t) \tag{3.39}$$

where $\tau_{d_i} > 0, 0 \leq |\varphi_{d_i}| < 1$, and

$$\hat{\omega}_{d_i}(t) = \gamma_{d_i} u_{n_i}(t) e^T(t) P B_i,\ \gamma_{d_i} > 0 \tag{3.40}$$

Remark 3.12: For the case of constant ideal weights in Assumption 3.1, Theorem 3.2 guarantees $e(t)$, $\tilde{W}(t)$, and $\hat{D}(t)$ approach a subspace in these error variables in which $e(t) = 0$ and $B[\hat{W}^T(t)\beta(x(t)) + \hat{D}u_n(t)] = 0$, similar to the result given by Theorem 3.2. Modification terms can also be employed, as described in Section 3.3.

Remark 3.13: In the case where $m > r$, control allocation can be used to permit some of the diagonal elements of Λ to be zero, as long as the pair $(A, B\Lambda)$ remains controllable. This is an important issue, particularly for flight control, which allows adaptive exploitation of redundancy in actuation in the event of actuator failure. The manner in which this extension can be implemented, as well as simulation results for several actuator failure modes applied to the NASA generic transport model (GTM), are illustrated in [10].

3.5 APPLICATION OF DF-MRAC TO FOUR SIMPLE EXAMPLES

In this section we begin by considering once again the two simple examples treated in Chapter 2, Section 2.3. We then include a third example illustrating the case where there are anomalies at the plant input and a fourth example illustrating the case of time delay at the plant input.

3.5.1 ANOMALIES WITH TIME-VARYING IDEAL WEIGHTS

Figure 2.2 in Chapter 2 shows the MRAC performance for the case of a first-order system with matched uncertainty in which the ideal weight used to represent the uncertainty is a square wave. Figure 3.3 shows the performance obtained for the same case in which the DF-MRAC adaptive law in Eq. (3.2) was used with $\Omega_1 = 0.5$, $\tau = 0.01$, and $\kappa_2 = 200$. Tracking performance is excellent, the control time history is reasonable, and the estimated weight is close to the ideal value except for a short transient period occurring at every step in the ideal weight. (The conditions for persistence of excitation are satisfied.) This suggests DF-MRAC can be highly advantageous in applications requiring upset recovery from a damage that suddenly alters the dynamic response of the system. This issue will be illustrated in several realistic aircraft flight control applications in Chapter 7.

As a variation of the previous example, the ideal weight history was changed to a sinusoidal function in which the frequency was varied from 0.5 rad/s up to 50 rad/s. The results are shown in Figs. 3.4–3.6. Figure 3.4 shows the MRAC controller is not able to cancel the uncertainty or estimate the ideal weight history at the lower setting for adaptation gain, and gives an even worse response when using the high setting for adaptation gain. Figures 3.5 and 3.6

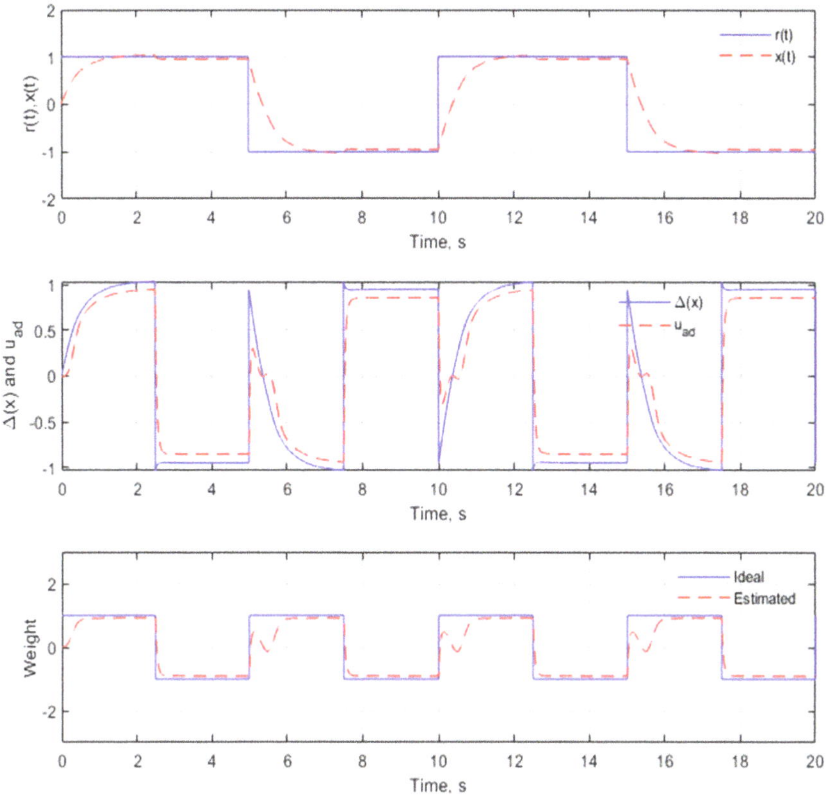

Fig. 3.3 DF-MRAC responses for the square wave ideal weight.

show that the DF-MRAC case for $\kappa_2 = 300$ gives an excellent response, and the estimated weight history is very close to the true weight, even at the high frequency end (see Fig. 3.6). Inspired by this result, we decided to try a case in which $w(t)$ is a band-limited white noise signal. The associated results are shown in Figs. 3.7 and 3.8. In particular, Fig. 3.8 shows the estimated weight is remarkably close to $w(t)$.

3.5.2 ANOMALIES WITH CONSTANT IDEAL WEIGHTS

Figures 2.3 and 2.4 in Chapter 2 show the MRAC performance for the case of a first-order system with matched uncertainty in which the ideal weight is constant, but the nominal controller has integral action. The performance of the MRAC controller was quite poor, suggesting the MRAC controller will not perform well when used to augment an existing P+I controller design. Figure 3.9 shows

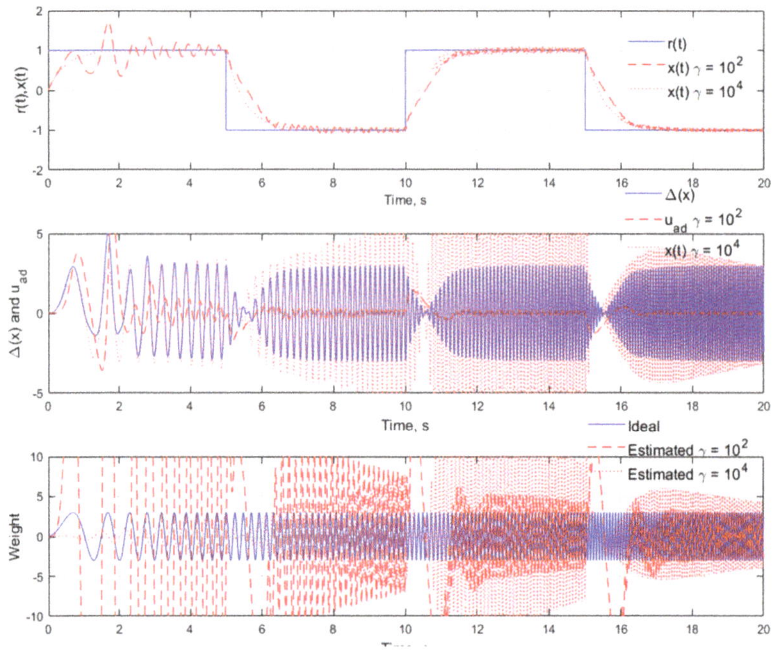

Fig. 3.4 MRAC responses for a sinusoidal ideal weight using $\gamma = 10^2$ and $\gamma = 10^4$.

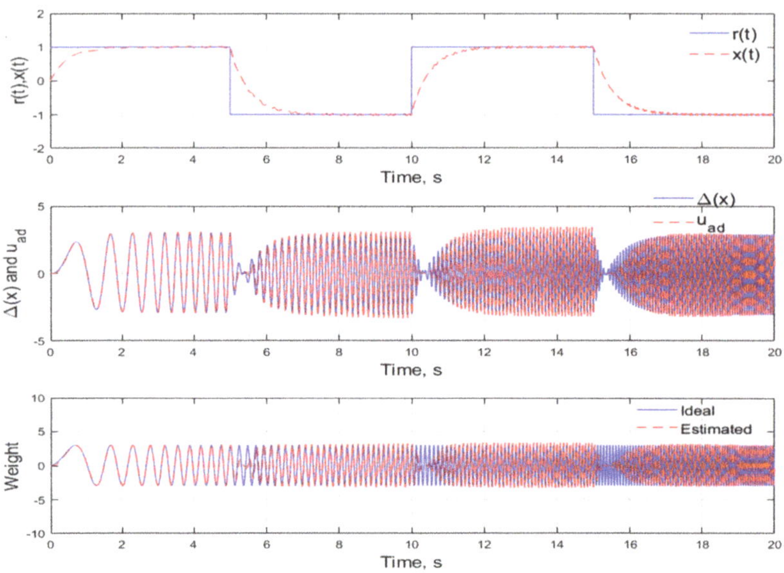

Fig. 3.5 DF-MRAC responses for a sinusoidal ideal weight using $\kappa_2 = 300$.

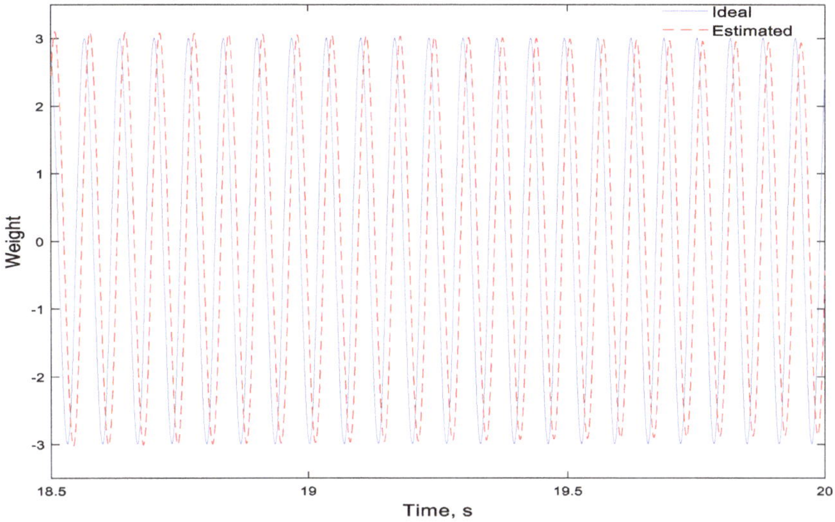

Fig. 3.6 Expanded view of the weight history at the high frequency end.

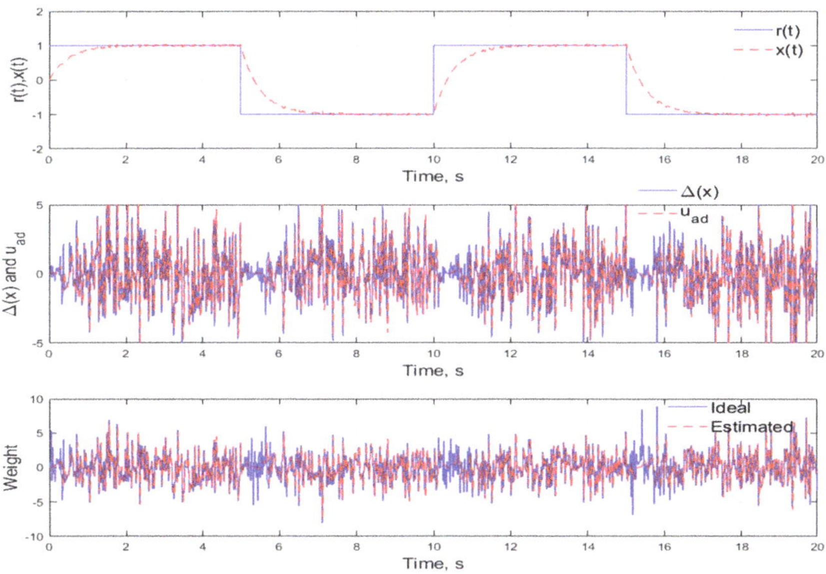

Fig. 3.7 DF-MRAC responses for a band-limited white noise ideal weight.

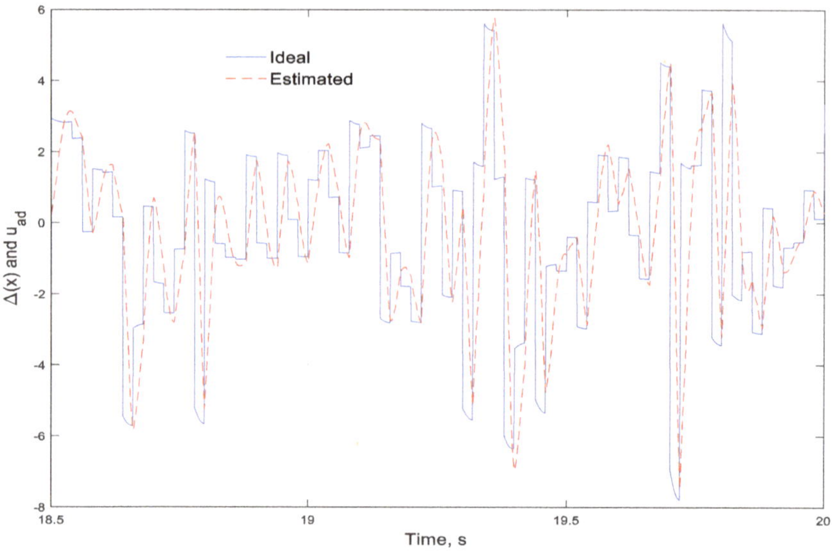

Fig. 3.8 Expanded view of the weight history near the end.

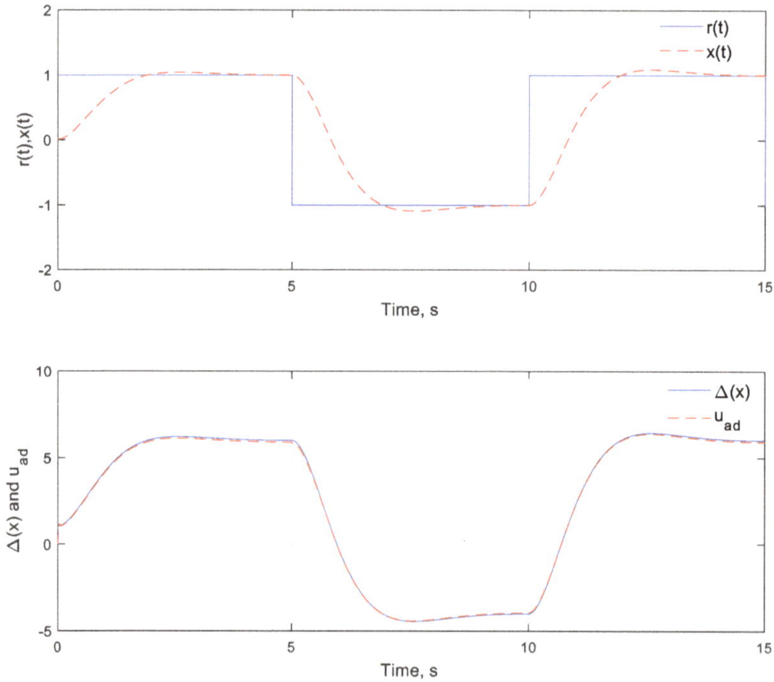

Fig. 3.9 DF-MRAC responses for a constant ideal weight.

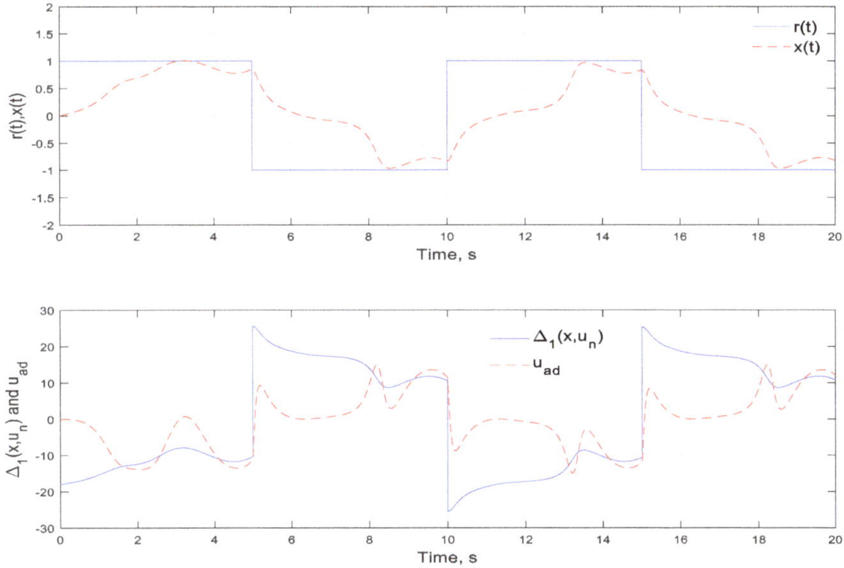

Fig. 3.10 DF-MRAC responses with input uncertainty when Eq. (3.2) is used for adaptation.

the performance obtained for the same case in Fig. 2.3 in which the DF-MRAC adaptive law in Eq. (3.2) was used with $\tau = 0.01$ and $\kappa_2 = 200$. Tracking performance is excellent, the control time history is reasonable, and the estimated uncertainty is very close to the true value. This result supports the statement made in Remark 3.9.

3.5.3 INPUT ANOMALIES

Here we examine DF-MRAC performance for a case containing matched uncertainty both in the plant dynamics and at the plant input. We consider the same first-order system treated previously in Section 3.5.1 using a sinusoidal ideal weight having an amplitude of 1.0 and a frequency of 2.5 rad/s to represent uncertainty in the plant dynamics. The plant input anomaly is represented using $\Lambda = 0.1$ in Eq. (3.34). Figure 3.10 shows the response obtained when the adaptive law in Eq. (3.2) is employed using the same parameter settings as in Section 3.5.1. Figure 3.11 shows the response obtained when the adaptive law in the form of Eq. (3.38) is employed using the same adaptation parameter settings.

3.5.4 TIME DELAY

We return to the example in Section 3.5.1 and introduce time delay to the case in which there is a square wave time variation in the ideal weight. This example illustrates the importance of ALR modification in adaptive control. Figure 3.12

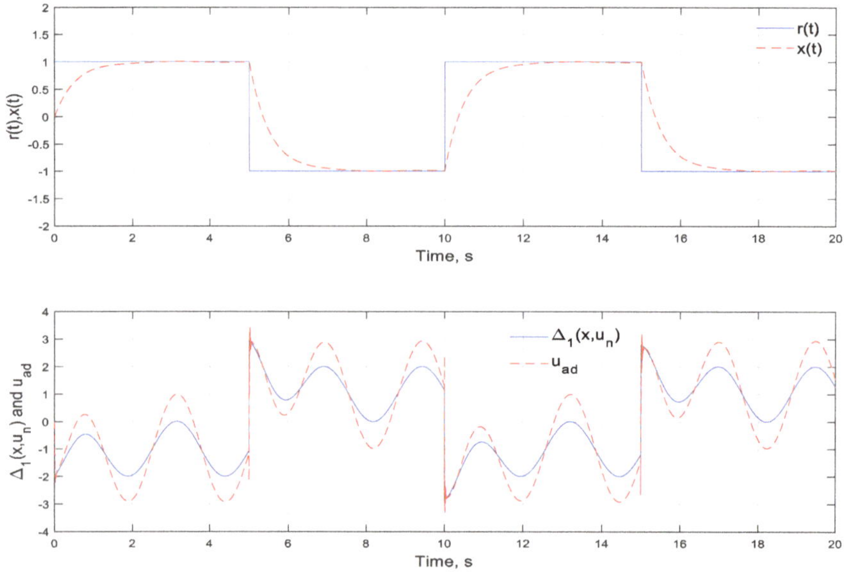

Fig. 3.11 DF-MRAC responses with input uncertainty using Eq. (3.38) for adaptation.

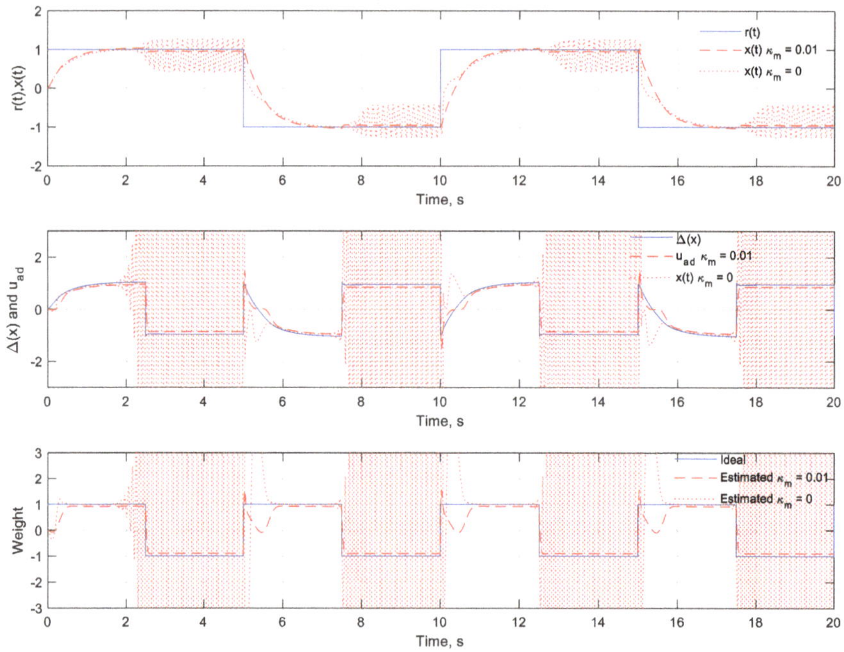

Fig. 3.12 DF-MRAC responses with 0.02-second time delay at the plant input.

shows that a small time delay of 0.02 seconds at the plant input has a profound effect on the time response, whereas the response using a modification gain of $\kappa_m = 0.01$ in Eq. (3.25) essentially recovers the performance shown in Fig. 3.3 obtained for the case without time delay. This result also shows adaptive control can be very sensitive to time delay. The response without adaptation to this level of time delay is nearly identical to the nominal control response in Fig. 2.1.

3.6 SUMMARY

The key feature of DF-MRAC is that the stability analysis is performed under the assumption the ideal weights are bounded, but otherwise arbitrarily time-varying. The approach is particularly useful for situations in which it is difficult or impossible to represent the uncertainty by a set of basis functions with unknown constant weights. The system error signals, including the state tracking error and the weight update error, are proven to be UUB using a Lyapunov–Krasovskii functional. An analysis is provided showing how the design parameters in the adaptive law can influence the ultimate bound and the exponential rate of convergence to the bound. The simple examples illustrate the superior performance of DF-MRAC in comparison with conventional MRAC, when it is evaluated for a variety of cases in which there is a sudden or rapidly varying set of uncertainties. DF-MRAC is shown to be particularly superior to MRAC when augmenting a nominal P+I controller. It also is shown how DF-MRAC can be modified using a variety of modification terms, and in particular one based on loop transfer recovery.

Chapter 7 presents aircraft simulation and flight test results to demonstrate the dramatic improvement in robustness to time delays achievable by using the derivative-free adaptation law with ALR modification. This is an important step toward certification of an adaptive flight control design because such designs can be much more sensitive to time delay and other unmodeled dynamic effects in comparison to nonadaptive designs.

REFERENCES

[1] Yucelen, T., and Calise, A. J., "Derivative-Free Model Reference Adaptive Control," *AIAA Journal of Guidance, Control, and Dynamics*, Vol. 34, No. 4, July–August 2011, pp. 933–950.

[2] Chen, W., and Chowdhury, F. N., "Simultaneous Identification of Time-Varying Parameters and Estimation of System States Using Iterative Learning Observers," *International Journal of Systems Science*, Vol. 38, No. 1, 2007, pp. 39–45.

[3] Hale, J., and Lunel, S. M. V., *Introduction to Functional Differential Equations*, Springer-Verlag, New York, 1993.

[4] Bernstein, D. S., *Matrix Mathematics: Theory, Facts, and Formulas*, 2nd ed., Princeton University Press, Princeton, NJ, 2009.

[5] Khalil, H. K., *Nonlinear Systems*, 2nd ed., Prentice Hall, Upper Saddle River, NJ, 1993.

[6] Slotine, J. J. E., and Li, W., *Applied Nonlinear Control*, Prentice Hall, Upper Saddle River, NJ, 1991.

[7] Narendra, K., and Annaswamy, A., "A New Adaptive Law for Robust Adaptation Without Persistent Excitation," *IEEE Transactions on Automatic Control*, Vol. 32, No. 2, 1987, pp. 134–145.

[8] Ioannou, P., and Kokotovic, P., "Instability Analysis and Improvement of Robustness of Adaptive Control," *Automatica*, Vol. 20, No. 5, 1984, pp. 583–594.

[9] Calise, A. J., and Yucelen, T., "Adaptive Loop Transfer Recovery," *AIAA Journal of Guidance, Control, and Dynamics*, Vol. 35, No. 3, May–June 2012, pp. 807–815.

[10] Yucelen, T., and Calise, A. J., "Adaptive Control for the Generic Transport Model: A Derivative-Free Approach," *AIAA InfoTech Conference*, Atlanta, GA, April 20–22, 2010.

CHAPTER 4

Robustness to Unmodeled Dynamics

4.1 THE ISSUE

Applications of adaptive control theory are challenged by the possible presence of unmodeled dynamics. In particular, because every physical system has unmodeled parasitic high-frequency dynamics, an adaptive design that does not take into account unmodeled dynamics may lead to an unstable closed-loop system [1–4]. This chapter examines the robustness of DF-MRAC to unmodeled dynamics. In contrast to derivative-based MRAC laws, this chapter shows that using a Lyapunov–Krasovskii functional [5] improves robustness to unmodeled dynamics in DF-MRAC by including a bias term in the set of basis functions. Derivative-based MRAC does not offer the same advantage. Several numerical examples are used to illustrate the advantages of a DF-MRAC within this context, including an application to a flexible spacecraft model.

4.2 PRELIMINARIES

In this section, a simplified adaptive control setting is considered to theoretically and numerically illustrate the fragile nature of derivative-based MRAC laws and the robust nature of DF-MRAC laws in the presence of unmodeled dynamics. Consider a second-order system given by

$$\dot{x}_p = -a_p x_p(t) + b_p u(t) + h_u x_u(t), \quad x_p(0) = x_{p0} \tag{4.1}$$

$$\dot{x}_u(t) = -a_u x_u(t) + h_p x_p(t), \quad x_u(0) = x_{u0} \tag{4.2}$$

where $x_p(t)$ is the modeled plant state, $x_u(t)$ is the unmodeled state, $u(t)$ is the control input, and a_p, b_p, h_u, a_u, and h_p are unknown parameters for which the sign of b_p is known and $a_u > 0$. Furthermore, consider a reference system satisfying a command tracking specification in terms of

$$\dot{x}_r(t) = -a_r x_r(t) + b_r r(t) \tag{4.3}$$

where $x_r(t)$ is the reference state, a_r and b_r are specified parameters with $a_r > 0$, and $r(t)$ is a bounded command. Equation (4.1) can be rewritten as

$$\dot{x}_p(t) = -a_r x_p(t) + b_r r(t) + b_p[u(t) + w_p^T \beta_p(x_p(t), r(t))] + h_u x_u(t) \tag{4.4}$$

where

$$w_p^T \triangleq \left[0,\ b_p^{-1}(a_r - a_p),\ -b_p^{-1} b_r\right] \tag{4.5}$$

is an unknown ideal weight and

$$\beta_p(x_p(t), r(t)) \triangleq [1, x_p(t), r(t)]^T \tag{4.6}$$

is a basis function that includes a bias term. The importance of including a bias term will become apparent later in Section 4.4.2.

4.2.1 DERIVATIVE-BASED MRAC

Taking into account x_u is unmodeled, consider the derivative-based MRAC law

$$u(t) = -\hat{w}_p^T \beta_p(x_p(t), r(t)) \tag{4.7}$$

$$\dot{\hat{w}}_p(t) = \gamma\, \mathrm{sign}(b_p) e(t) \beta_p(x_p(t), r(t)), \quad \hat{w}_p(0) = \hat{w}_{p0} \tag{4.8}$$

where

$$e(t) \triangleq x_p(t) - x_r(t) \tag{4.9}$$

is the tracking error, and $\hat{w}_p(t)$ is an estimate of w_p satisfying the update law in Eq. (4.8) with adaptation gain $\gamma > 0$. The stability properties of the system given by Eqs. (4.2), (4.4), (4.7), and (4.8) can be examined by considering the Lyapunov function candidate

$$L(e(t), \tilde{w}_p(t), x_u(t)) = e^2(t)/2 + x_u^2(t)/2 + \gamma^{-1}|b_p| \tilde{w}_p^T(t) \tilde{w}_u(t)/2 \tag{4.10}$$

where

$$\tilde{w}_p(t) \triangleq w_p - \hat{w}_p(t) \tag{4.11}$$

is the weight estimation error, and $e(t)$ satisfies

$$\dot{e}(t) = -a_r e(t) + b_p \tilde{w}_p^T(t) \beta_p(x_p(t), r(t)) + h_u x_u(t) \tag{4.12}$$

Differentiation of Eq. (4.10) yields

$$\dot{L}(e(t), \tilde{w}_p(t), x_u(t)) \le -\begin{bmatrix} |e(t)| \\ x_u(t) \end{bmatrix}^T \times \begin{bmatrix} a_r & -(|h_p| + |h_u|)/2 \\ -(|h_p| + |h_u|)/2 & a_u \end{bmatrix} \begin{bmatrix} |e(t)| \\ |x_u(t)| \end{bmatrix} + |h_p||x_r(t)||x_u(t)| \tag{4.13}$$

Then

$$\begin{bmatrix} a_r & -(|h_p| + |h_u|)/2 \\ -(|h_p| + |h_u|)/2 & a_u \end{bmatrix} > 0 \tag{4.14}$$

or

$$a_r a_u > (|h_p| + |h_u|)^2/4 \tag{4.15}$$

is sufficient to ensure closed-loop stability, because the last term in Eq. (4.13) is linear in $x_u(t)$, and $r(t)$ is bounded. This illustrates a limitation of the derivative-based weight adaptation law in Eq. (4.8) because the only way to improve robustness to unmodeled dynamics is to increase a_r. However, a_r is used to define the desired performance of the closed-loop system, and as such should not be considered a parameter that can be chosen sufficiently large in order to satisfy Eq. (4.15). For example, increasing a_r tends to increase sensitivity to sensor noise and the likelihood of control saturation. Using standard modification terms in the update law of Eq. (4.8) such as σ-modification [6], e-modification [7], or a projection operator [8] leads to a condition similar to Eq. (4.14) and to the same conclusion

4.2.2 DF-MRAC

Instead of Eq. (4.8), consider the DF-MRAC update law given by

$$\hat{w}_p(t) = \gamma_1 \hat{w}_p(t-\tau) + \gamma_2 \mathrm{sign}(b_p) e(t) \beta_p(x_p(t), r(t)) \tag{4.16}$$

where $\tau > 0$, $0 < \gamma_1 < 1$, and $\gamma_2 > 0$ is the adaptation gain. Next it is shown that increasing γ_2 improves robustness to unmodeled dynamics.

The stability properties of the system given by Eqs. (4.1), (4.2), (4.7), and (4.16) can be examined by considering the Lyapunov–Krasovskii functional

$$L_k(e(t), \tilde{w}_{pt}, x_u(t)) = e^2(t)/2 + x_u^2(t)/2 + \rho|b_p| \int_{t-\tau}^{t} \tilde{w}_p^T(s)\tilde{w}_p(s)ds/2 \tag{4.17}$$

where $\tilde{w}_{pt}$ denotes $\tilde{w}_p(t)$ over the interval from $t-\tau$ to t. Differentiation of Eq. (4.17) yields

$$\dot{L}_k(e(t), \tilde{w}_{pt}, x_u(t)) = e(t)\dot{e}(t) + x_u \dot{x}_u + \rho|b_p|[\tilde{w}_p^T(t)\tilde{w}_p(t) - \tilde{w}_p^T(t-\tau)\tilde{w}_p(t-\tau)]/2 \tag{4.18}$$

Express $\tilde{w}_p(t)$ in the form

$$\begin{aligned}\tilde{w}_p(t) &= \gamma_1 \tilde{w}_p(t-\tau) + \Omega - \gamma_2 \mathrm{sign}(b_p)\beta_p(x_p(t), r(t))e(t),\\ \Omega &\triangleq (1-\gamma_1)w_p\end{aligned} \tag{4.19}$$

Using Eqs. (4.2), (4.12), and (4.19) in Eq. (4.18) and selecting $\rho = 1/\gamma_2$ [similar to what was done in Eq. (3.8) in Chapter 3], it follows that

$$\begin{aligned}&\dot{L}_k(e(t), \tilde{w}_p(t), x_u(t)) = -a_r e^2(t) - a_u x_u^2(t)\\ &-\gamma_2 |b_p| e^2(t)\beta_p^T(\cdot)\beta_p(\cdot)/2 - |b_p|\left[(1-\gamma_1^2)\tilde{w}_p^T(t-\tau)\tilde{w}_p(t-\tau) + \Omega^T\Omega\right]/2\gamma_2\\ &+ h_u x_u(t)e(t) + h_p x_u(t)[e(t)+x_u(t)] + |b_p|\gamma_1\Omega^T\tilde{w}_p(t-\tau)/\gamma_2\\ &\le -(a_r + \gamma_2|b_p|/2)e^2(t) - a_u x_u^2(t)\\ &-|b_p|\left[(1-\gamma_1^2)\tilde{w}_p^T(t-\tau)\tilde{w}_p(t-\tau) + \Omega^T\Omega\right]/2\gamma_2 + (h_p + h_u)x_u(t)e(t)\\ &+ h_p x_u(t)x_r(t) + |b_p|\gamma_1\Omega^T\tilde{w}_p(t-\tau)/\gamma_2\end{aligned} \tag{4.20}$$

where $\beta_p(\cdot)$ in the equality in Eq. (4.20) denotes the dependence of β_p on $x_p(t)$ and $r(t)$ as defined in Eq. (4.6). Note the inequality in Eq. (4.20) follows from the fact that $\beta_p^T(\cdot)\beta_p(\cdot) \ge 1$, which is a direct consequence of our employing a bias term in the set of basis functions.

Next consider Young's inequality [9]

$$2x^T y \le c x^T x + y^T y/c, \quad c > 0 \tag{4.21}$$

where x and y are vectors. Applying this to the last term in Eq. (4.20) in the form

$$\Omega^T \tilde{w}_p(t-\tau) \le c\Omega^T\Omega/2 + \tilde{w}_p^T(t-\tau)\tilde{w}_p(t-\tau)/2c \tag{4.22}$$

it follows from Eq. (4.20) that

$$\begin{aligned}\dot{L}_k(e(t), x_u(t)) \le &-\begin{bmatrix}|e(t)|\\ |x_u(t)|\end{bmatrix}^T \begin{bmatrix} a_r + \gamma_2|b_p|/2 & -(|h_p| + |h_u|)/2\\ -(|h_p|+|h_u|)/2 & a_u\end{bmatrix}\begin{bmatrix}|e(t)|\\ |x_u(t)|\end{bmatrix}\\ &-|b_p|\left[\tilde{w}_p^T(t)\tilde{w}_p(t) + (1-(1+1/c\gamma_1)\gamma_1^2)\tilde{w}_p^T(t-\tau)\tilde{w}_p(t-\tau)\right]/2\gamma_2\\ &+ |h_p||x_r(t)||x_u(t)| + |b_p|(1+c\gamma_1)\Omega^T\Omega/2\gamma_2\end{aligned} \tag{4.23}$$

Choosing $c > \gamma_1/(1-\gamma_1^2)$ ensures $(1-(1+1/c\gamma_1)\gamma_1^2) > 0$. The third term in Eq. (4.23) is linear in $|x_u(t)|$ and $x_r(t)$ is bounded. The last term in

Eq. (4.23) is also bounded. Therefore,

$$\left(a_r + \gamma_2 \left|b_p\right|/2\right) a_u > \left(\left|h_p\right| + \left|h_u\right|\right)^2/4 \tag{4.24}$$

is sufficient to ensure $z(t) = \left[e(t), \tilde{w}_p(t), x_u(t)\right]^T$ is UUB.

Remark 4.1: In contrast to the condition in Eq. (4.15), the condition in Eq. (4.24) implies robustness of the adaptive system to unmodeled dynamics can be improved by increasing γ_2. The fact that the condition in Eq. (4.24) depends on γ_2 is a direct consequence of including a bias term in the basis vector.

Remark 4.2: When we assume the parameterization of uncertainty is exact, as in Eqs. (4.5) and (4.6), then theoretically there is no limit on the adaptation gain. However, if we assume the parameterization is exact on a bounded set, this will, in turn, imply a practical upper limit on the adaptation gain such that the tracking error is guaranteed to be UUB. In [10] we assume the latter and derive an expression for the limit on γ_2. (See in particular Remark 5 in [10].)

4.2.3 A NUMERICAL EXAMPLE

In this section we compare the performance of derivative-based MRAC [using Eq. (4.8) as the weight update law] with DF-MRAC [using Eq. (4.16) as the weight update law]. Let $a_p = -1$, $b_p = 1$, $a_r = 0.5$, $b_r = 0.5$, and $r(t) = \sin(0.5t)$ in Eqs. (4.1) and (4.3). Furthermore, assume Eq. (4.1) is coupled with a second-order, nonminimum phase, lightly damped system described by

$$\dot{x}_u(t) = A_u x_u + \overline{h}_p x_p(t) \tag{4.25}$$

$$y_u(t) = \overline{h}_u x_u(t) \tag{4.26}$$

$$A_u = \begin{bmatrix} -0.001 & -1 \\ 1 & 0 \end{bmatrix}, \quad \overline{h}_p = \begin{bmatrix} 4 \\ 0 \end{bmatrix}, \quad \overline{h}_u = \begin{bmatrix} 1.25 & 6.25 \end{bmatrix} \tag{4.27}$$

Remark 4.3: The preceding analysis is extended to this case by replacing $x_u^2(t)/2$ in Eqs. (4.10) and (4.17) with $x_u^T S x_u(t)$, where S satisfies $0 = A_u^T S + S A_u + Q$, $Q > 0$. If we let $Q = I$, then the sufficient condition in the case of derivative-based MRAC in Eq. (4.15) becomes

$$a_r - \left(\left|h_u\right|/2 + \overline{\sigma}(S)\left|h_p\right|\right)^2 > 0 \tag{4.28}$$

and the sufficient condition in the case of DF-MRAC in Eq. (4.24) becomes

$$a_r + \gamma_2\left|b_p\right|/2 - \left(\left|h_u\right|/2 + \overline{\sigma}(S)\left|h_p\right|\right)^2 > 0 \tag{4.29}$$

Although $\overline{\sigma}(S)$ is not directly relatable to a bound on A_u, it can be seen that increasing γ_2 will improve robustness to unmodeled dynamics.

The adaptation gain is set to $\gamma = 25$ for derivative-based MRAC. For DF-MRAC we used $\tau = 0.025$, $\gamma_1 = 0.5$, and $\gamma_2 = 25$. The control laws are

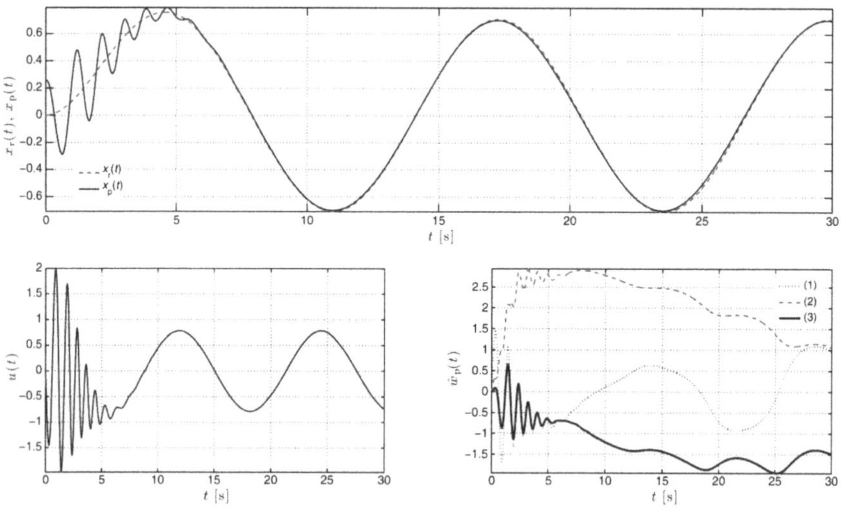

Fig. 4.1 Derivative-based MRAC performance without unmodeled dynamics ($\gamma = 25$).

implemented in discrete time, using a sample rate of 200 Hz. Figure 4.1 shows the derivative-based MRAC response without unmodeled dynamics, for an initial condition $x_p(0) = 0.25$. Tracking performance $x_p(t) - r(t)$ is somewhat poor during the transient phase, which lasts for approximately 5 seconds. This cannot be improved by simply increasing the adaptation gain. Figure 4.2 shows

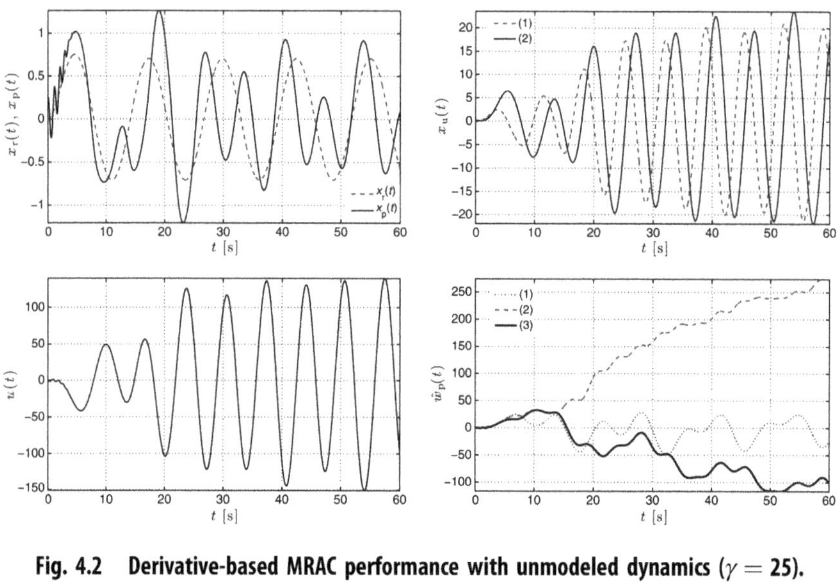

Fig. 4.2 Derivative-based MRAC performance with unmodeled dynamics ($\gamma = 25$).

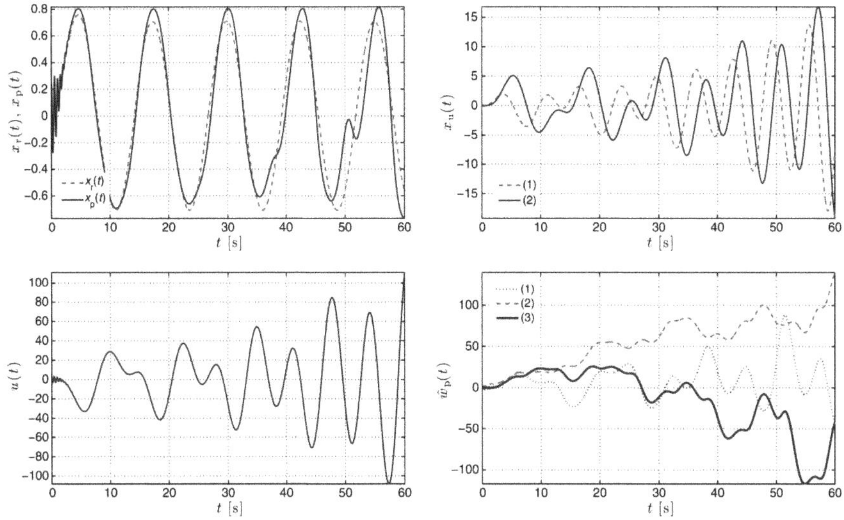

Fig. 4.3 Derivative-based MRAC performance with unmodeled dynamics ($\gamma = 125$).

tracking performance is very poor when unmodeled dynamics are included. Figure 4.3 shows the behavior when the adaptation gain is increased to $\gamma = 125$. Tracking performance initially improves, but the system eventually goes unstable (not shown). Figures 4.4 and 4.5 show what happens when

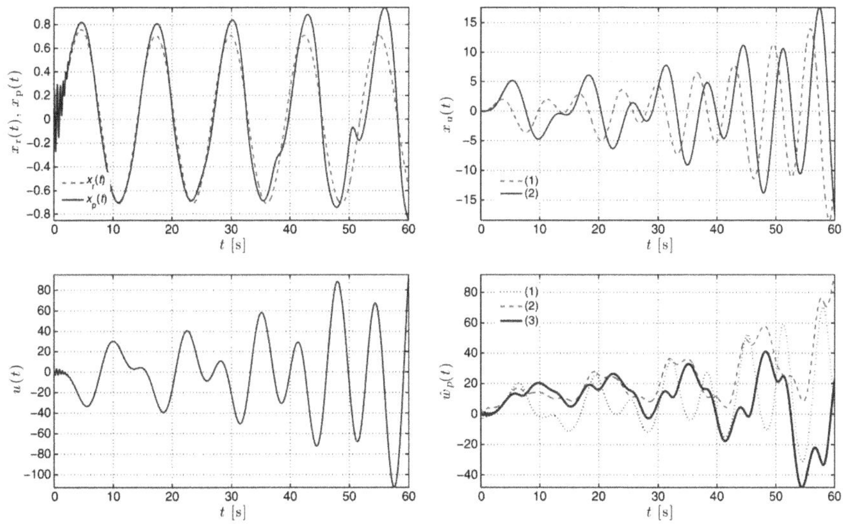

Fig. 4.4 Derivative-based MRAC with σ-modification with unmodeled dynamics.

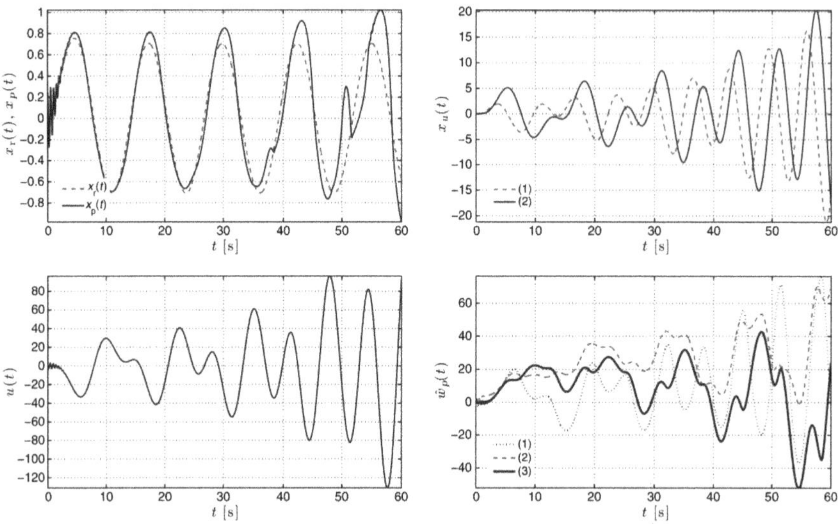

Fig. 4.5 Derivative-based MRAC with *e*-modification with unmodeled dynamics.

σ- and e-modification terms are added to the case in Fig. 4.3. The responses get worse if the gains on the modification terms are increased. A similar result is obtained when parameter projection [11] is used to limit the growth in the adaptation gain, and the rapidity of the instability in the state and control responses increases as the limit on the norm of the weight vector is reduced.

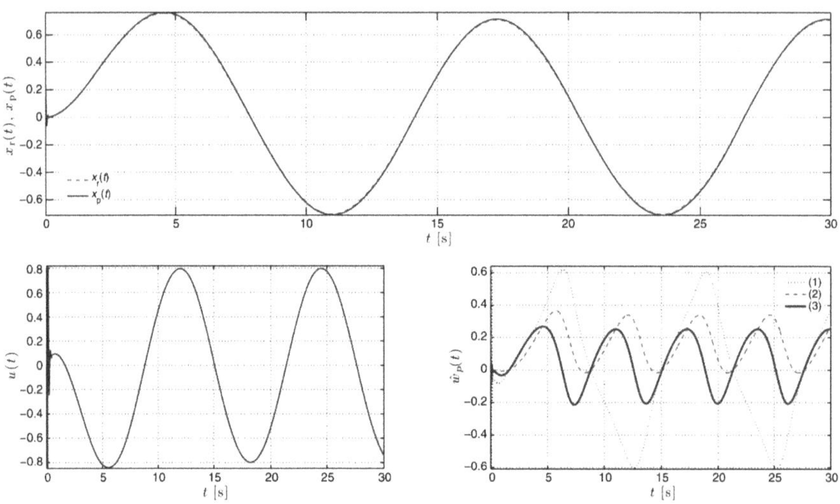

Fig. 4.6 DF-MRAC without unmodeled dynamics ($\tau = 0.025$, $\gamma_1 = 0.5$, $\gamma_2 = 25$).

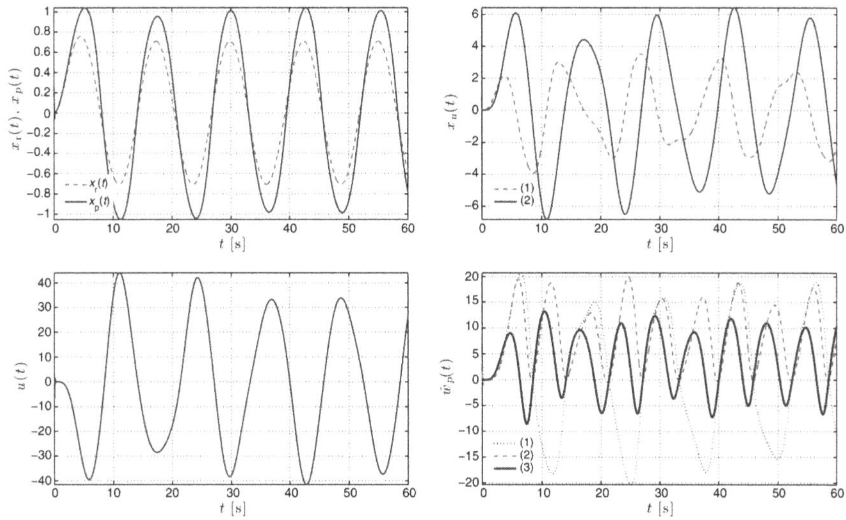

Fig. 4.7 DF-MRAC with unmodeled dynamics ($\tau = 0.025$, $\gamma_1 = 0.5$, $\gamma_2 = 25$).

Figure 4.6 shows the DF-MRAC response without unmodeled dynamics. Note the initial transient in the tracking error is imperceptible, a result consistent with the claims made in Chapter 3. Figure 4.7 shows the response with unmodeled dynamics. The tracking performance is significantly improved in comparison to the response in Fig. 4.2. Moreover, the responses of the unmodeled states are bounded to much smaller values than the responses in Fig. 4.2, and the corresponding level of control effort is dramatically reduced.

4.3 THE MULTI-INPUT, MULTISTATE CASE

Consider the uncertain system

$$\dot{x}_p(t) = A_p x_p(t) + B_p \left[\Lambda u(t) + W_p^T(t) \beta_p(x_p(t)) + W_u^T(t) x_u(t) \right], \quad x_p(0) = x_{P0} \tag{4.30}$$

$$\dot{x}_u(t) = A_u x_u(t) + B_u x_p(t), \quad x_u(0) = x_{u0} \tag{4.31}$$

where $x_p(t) \in R^{n_p}$ is the state vector associated with the modeled portion of the system dynamics, which is assumed to be accessible for feedback; $x_u(t) \in R^{n_u}$ contains the elements of the state vector associated with the unmodeled dynamics, for which n_u is also unknown; $u(t) \in R^m$ is the control input; $A_p \in R^{n_p x n_p}$ and $B_p \in R^{n_p x m}$ are known system matrices; $\Lambda \in D_+^{mxm}$ is an unknown control effectiveness matrix; $W_p \in R^{(n_p+1)xm}$ is an unknown but bounded ideal weight matrix; $\beta_p : R^{n_p} \to R^{n_p+1}$ is a known basis function of the form $\beta_p \triangleq \left[1, x_p^T(t)\right]^T$; and $A_u \in R^{n_u x n_u}$ and $B_u \in R^{n_u x n_p}$ are unknown system matrices associated with the unmodeled dynamics. Note the first element of β_p is a bias term, similar to what was done in Eq. (4.6).

Assumption 4.1: The pair (A_p, B_p) is controllable, A_u is Hurwitz, and $W_p(t)$ and $W_u(t)$ are bounded (i.e., there exists w_p^* and w_u^* such that $\|W_p(t)\| \leq w_p^*$ and $\|W_p(t)\| \leq w_u^*$).

Remark 4.4: Because A_u is Hurwitz, the Lyapunov equation

$$0 = A_u^T S + S A_u + Q \tag{4.32}$$

with $Q \in R_+^{n_u x n_u}$ has a unique solution positive definite solution for S, although it is not computable because A_u is otherwise unknown.

Remark 4.5: The condition $\Lambda \in D_+^{mxm}$ on the unknown control effectiveness matrix can be relaxed to $\Lambda \in \overline{D}_+^{mxm}$ if the range space of B_p is equal to the range space of $B_p\Lambda$. This implies actuator redundancy exists. An actuator is considered redundant if its failure does not change the range space of $B_p\Lambda$. This is often true in aircraft applications, where the means of control is primarily through moment generation (see the generic transport model (GTM) example in Chapter 7).

Next consider the feedback control law

$$u(t) = u_n(t) + u_a(t) \tag{4.33}$$

where $u_n(t) \in R^m$ and $u_a(t) \in R^m$ are the nominal and adaptive control laws. The nominal control law has a standard proportional control form

$$u_n(t) = -K_1 x_p(t) + K_2 r(t), \quad K_1 \in R^{mxn_p}, \quad K_2 \in R^{mxr} \tag{4.34}$$

where $r(t) \in R^r$ is a bounded piecewise continuous command input and $A_m \triangleq A_p - BK_1$ is Hurwitz. (The nominal controller ensures the nominal closed loop system is stable.)

Remark 4.6: The extension to nominal control in P+I form can be carried out in a manner similar to what was done in Chapter 2 by augmenting $x_p(t)$ with integrator states.

Using Eqs. (4.32) and (4.33) in Eq. (4.30) yields

$$\begin{aligned} \dot{x}_p(t) &= A_m x_p(t) + B_p\Lambda[K_2 r(t) + u_a(t) \\ &\quad + W^T(t)\beta(x_p(t))] + B_p W_u^T(t) x_u(t) \end{aligned} \tag{4.35}$$

where $W^T(t) \triangleq [\Lambda^{-1} W_p^T(t), (\Lambda^{-1} - I_{mxm})K_1] \in R^{(2n_p+1)xm}$ is an unknown (aggregated) weight matrix and $\beta^T(x_p(t)) \triangleq \left[\sigma_p^T(x_p(t)), x_p^T(t)\right], \in R^{(2n_p+1)}$ is a known (aggregated) basis function. Considering Eq. (4.35), let the adaptive control law be

$$u_a(t) = -\hat{W}^T(t)\beta\big(x_p(t)\big) \tag{4.36}$$

where $\hat{W}^T(t) \in R^{2n_p+1}$ is an estimate of W satisfying the DF update law

$$\hat{W}(t) = \gamma_1 \hat{W}(t-\tau) + \gamma_2\, \beta(x_p(t))e^T(t)PB_p \tag{4.37}$$

$0 < |\gamma_1| < 1,\ \gamma_2 \in R_+$ is the adaptation gain, $e(t) \triangleq x_p(t) - x_m(t)$ is the system error, with $x_m(t)\ \in R^{n_p}$ being the reference state vector, satisfying

$$\dot{x}_m(t) = A_m x_m(t) + B_m r(t),\quad x_m(t) = x_{m_0} \tag{4.38}$$

where $B_m = B_p\Lambda K_2$ and $P \in R_+^{n_p x n_p}$ is a solution of the Lyapunov equation

$$0 = A_m^T P + PA_m + R,\quad R \in R_+^{n_p x n_p} \tag{4.39}$$

Combining Eqs. (4.34), (4.35), and (4.37), the error dynamics can be written in the form

$$\dot{e}(t) = A_m e(t) + B_p\Lambda\tilde{W}^T(t)\beta(x_p(t)) + B_p W_u^T x_u(t),\quad e(0) = e_0 \tag{4.40}$$

where $e_0 \triangleq x_{p_o} - x_{m_0}$ and $\tilde{W}(t) \triangleq W(t) - \hat{W}(t)$ is the weight error, satisfying

$$\tilde{W}(t) = \gamma_1 \tilde{W}(t-\tau) + \Omega(t) - \gamma_2\sigma(x_p(t))e^T(t)PB_p \tag{4.41}$$

with

$$\Omega(t) \triangleq W(t) - \gamma_1 W(t-\tau) \tag{4.42}$$

The following theorem states the main robustness result of DF-MRAC:

Theorem 4.1: Consider the uncertain system defined by Eqs. (4.30) and (4.31), subject to Assumption 4.1. Consider, in addition, the feedback control law given by Eq. (4.33), with the nominal feedback control component given by Eq. (4.34) and the adaptive feedback control component given by Eq. (4.36), which has a derivative-free weight update law in the form of Eq. (4.37). Then the triple $[e(t),\ x_u(t),\ \tilde{W}(t)]$ satisfying Eqs. (4.40), (4.31), and (4.41), respectively, is UUB if

$$\underline{\lambda}(\Lambda) - k_1\gamma_2^{-1} \geq 0 \tag{4.43}$$

$$1 - k_2\underline{\lambda}^{-1}(R) > 0 \tag{4.44}$$

$$\underline{\lambda}(Q) - w_u^*/k_1 - \overline{\sigma}^2(SB_u)/k_2 > 0 \tag{4.45}$$

all hold for some subset of values for $k_1 > 0$ and $k_2 > 0$.

Proof: The proof of Theorem 4.1 parallels the line of arguments given in Section 4.2.2. The reader is referred to the Appendix of [9] for the details.

4.4 APPLICATION TO A FLEXIBLE SPACECRAFT MODEL

Consider the spacecraft model with flexible dynamics [12] given by:

$$J\dot{\omega} + \omega(t) \times \left[J\omega(t) + \delta^T \dot{\eta}\right] + \delta^T \ddot{\eta}(t) = u(t), \quad \omega(0) = \omega_0 \tag{4.46}$$

$$\ddot{\eta}(t) + C\dot{\eta}(t) + K\eta + \delta\dot{\omega}(t) = 0, \quad \dot{\eta}(0) = \dot{\eta}_0, \quad \eta(0) = \eta_0 \tag{4.47}$$

where $J \in R^{3\times 3}$ is the inertia matrix, $\omega \in R^3$ is the angular velocity with respect to an inertial frame expressed in the body frame, $\eta \in R^N$ is the modal coordinate vector relative to the main body, $u \in R^3$ is the control input vector, $\delta \in R^{Nx3}$ is the coupling matrix between the spacecraft's rigid and flexible dynamics, $K = \text{diag}\left(\Lambda_1^2, \ldots .\Lambda_N^2\right) \in R^{NxN}$ is the stiffness matrix, $C = \text{diag}(2\xi_1\Lambda_1, \ldots .2\xi_N\Lambda_N) \in R^{NxN}$ is the damping matrix, and Λ_i and ξ_i are the natural frequency and damping ratio of the ith flexible mode, respectively. The attitude kinematics are given by

$$\dot{q}(t) = [q(t) \times \omega(t) + \overline{q}(t)\omega(t)]/2, \; q(0) = q_0 \tag{4.48}$$

$$\dot{\overline{q}}(t) = q^T\omega(t)/2, \; q(0) = q_0 \tag{4.49}$$

where $\left[\overline{q}, q^T\right] \in R^4$ denotes the unit quaternion vector. The parameter values in Eqs. (4.46) and (4.47) are

$$J = \begin{bmatrix} 350 & 3 & 4 \\ 3 & 270 & 10 \\ 4 & 10 & 190 \end{bmatrix} \text{kg} \cdot \text{m}^2 \tag{4.50}$$

$$\delta = \begin{bmatrix} 6.46 & 1.28 & 2.16 \\ -1.26 & 0.918 & -1.67 \\ 1.12 & 2.49 & -0.837 \end{bmatrix} \text{kg}^{1/2} \cdot \text{m} \tag{4.51}$$

$N=3$, $\Lambda_1=0.768$, $\Lambda_2=1.10$, $\Lambda_3=1.87$, $\xi_1=0.0056$, $\xi_2=0.0086$, and $\xi_3=0.013$.

Linearization of Eqs. (4.46) and (4.47) results in an unmatched uncertainty due to the $q \times \omega$ term in Eq. (4.47). However, by defining $z_1 \triangleq q$ and $z_2 \triangleq 2\dot{q}$, Eqs. (4.46) and (4.47) can be written as

$$\dot{z}_1(t) = z_2(t)/2 \tag{4.52}$$

$$\dot{z}_2(t) = \overline{Q}J_0^{-1}u(t) + G[q(t), \; \omega(t), \; \eta(t), \; \dot{\eta}(t)] \tag{4.53}$$

where $J_0 \triangleq J - \delta^T\delta$ and

$$\overline{Q} = \begin{bmatrix} \overline{q} & -q_3 & q_2 \\ q_3 & \overline{q} & -q_1 \\ -q_2 & -q_1 & \overline{q} \end{bmatrix} \tag{4.54}$$

If this system is linearized about the equilibrium attitude $\overline{q}_0 = 1$ and $q = [0, 0, 0]^T$, and $\delta = 0_3$ is assumed, then in equilibrium $\overline{Q} = I_3$ and A_p and B_p in Eq. (4.30) become

$$A_p = \begin{bmatrix} 0_3 & 0.5I_3 \\ 0_3 & 0_3 \end{bmatrix}, \quad B_p = \begin{bmatrix} 0_3 \\ J^{-1} \end{bmatrix} \tag{4.55}$$

The nominal control law has the proportional form

$$u_n(t) = -K_1 \left[q^T(t), \omega^T(t)\right]^T + K_2 q_c(t) \tag{4.56}$$

where linear quadratic regulator theory was used to design K_1 using $Q = I_6$ to penalize $[q^T(t), \omega^T(t)]$ and $R = 0.001I_3$ to penalize $u(t)$, and selecting $K_2 = -\left[(I_3, 0_3)A_m^{-1}B_p\right]^{-1}$ so $q_m(t)$ tracks q_c with zero steady-state error in response to a step command, where $q_m(t)$ is the attitude response of the reference model. The reference system has the form in Eq. (4.38) with $B_r = B_pK_2$. The adaptive feedback control law has the form

$$u_a(t) = -\hat{W}^T(t)\beta\left[q^T(t), \omega^T(t)\right] \tag{4.57}$$

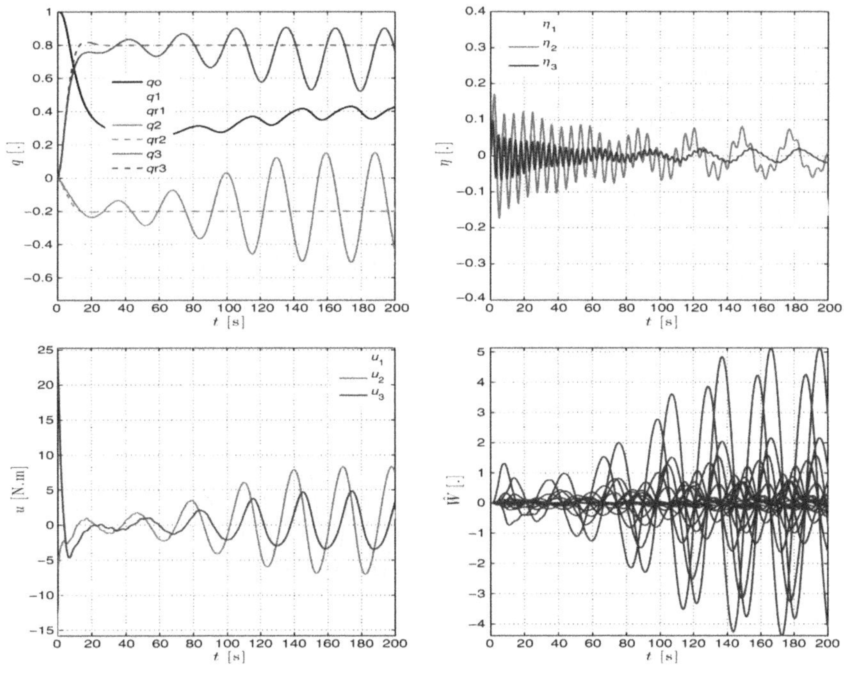

Fig. 4.8 Derivative-based MRAC responses for a flexible spacecraft model.

where $\hat{W}(t)$ satisfies the derivative-free weight update law given by Eq. (4.37) and the set of basis functions is chosen as

$$\beta\{q^T(t), \omega^T(t)\} = \left[1, \frac{1-e^{\omega_1}}{1+e^{\omega_1}}, \frac{1-e^{\omega_2}}{1+e^{\omega_2}}, \frac{1-e^{\omega_3}}{1+e^{\omega_3}}, \frac{1-e^{q_1}}{1+e^{q_1}}, \frac{1-e^{q_2}}{1+e^{q_2}}, \frac{1-e^{q_3}}{1+e^{q_3}}\right]^T \tag{4.58}$$

Figure 4.8 shows that the derivative-based MRAC system response to $q_c = [-0.5, -0.2, 0.6]^T$, with zero initial conditions, an adaptation gain of 1000, and a σ-modification gain of 0.5, is unstable. This is generally the case regardless of how these design parameters are chosen, and the same is true with e-modification and parameter projection (and combinations thereof). Figures 4.9 to 4.11 show the DF-MRAC system responses using $\tau = 0.1$, $\gamma_1 = 0.9$, and $\gamma_2 = 100$, 500, and 1000, respectively, and a control update rate of 400 Hz. These figures illustrate how robustness to unmodeled dynamics is improved as the adaptation gain is increased without the use of a modification term. Note the case $\gamma_2 = 100$ produces an unstable response and the responses become increasingly more stable as the adaptation gain is increased, to better satisfy the condition in Eq. (4.43).

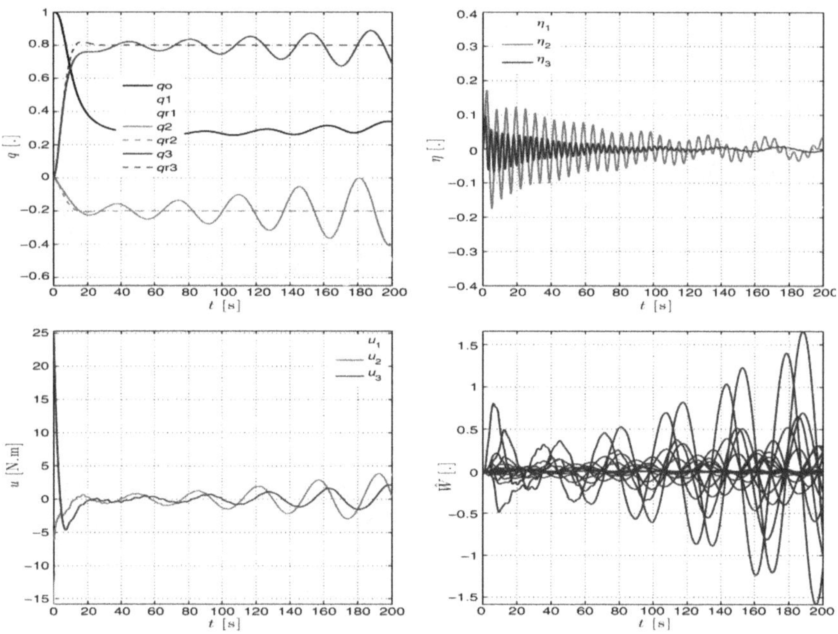

Fig. 4.9 DF-MRAC responses for a flexible spacecraft model using $\gamma_2 = 100$.

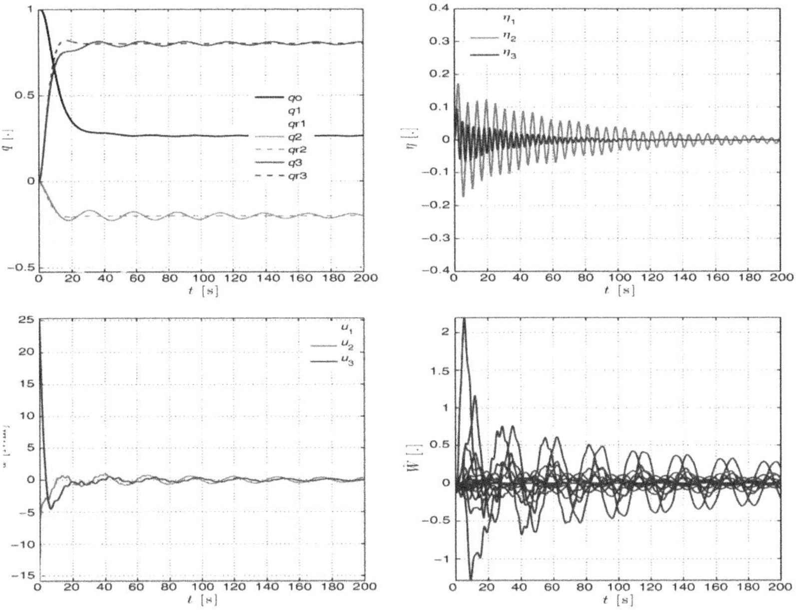

Fig. 4.10 DF-MRAC responses for a flexible spacecraft model using $\gamma_2 = 500$.

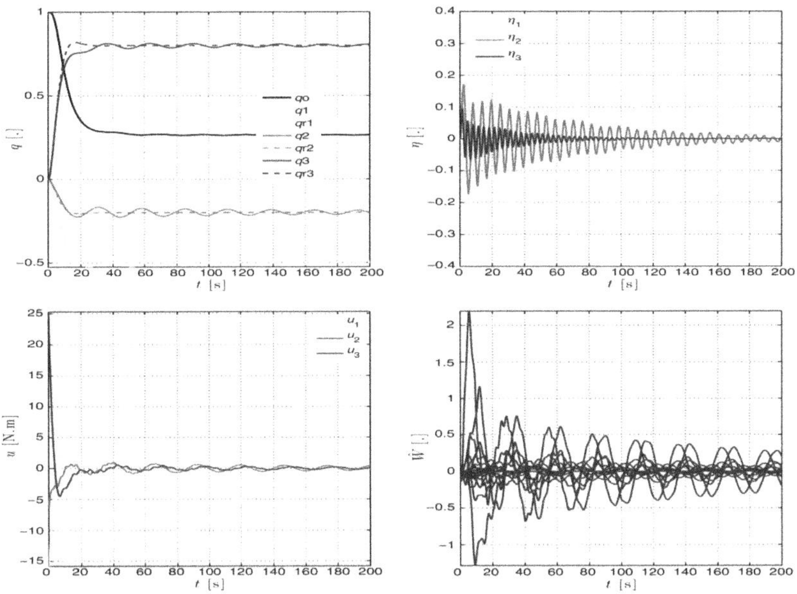

Fig. 4.11 DF-MRAC responses for a flexible spacecraft model using $\gamma_2 = 1000$.

4.5 SUMMARY

This chapter has shown that, in contrast to derivative-based MRAC, DF-MRAC is inherently robust to unmodeled dynamics, so long as a bias term is included in the set of basis functions. Furthermore, robustness can be improved by simply increasing the adaptation gain, without the need for modification terms. Derivative-based MRAC laws do not offer the same advantage. The examples provided illustrate the superior performance of DF-MRAC with reduced levels of control effort in comparison with a derivative-based MRAC adaptive law in the presence of system uncertainties and unmodeled dynamics.

REFERENCES

[1] Rohrs, C. E., Valavani, L. S., Athans, M., and Stein, G. "Robustness of Continuous-Time Adaptive Control Algorithms in the Presence of Unmodeled Dynamics," *IEEE Transactions on Automatic Control*, Vol. 30, 1985, pp. 881–889.

[2] Middleton, R.H, Goodwin, G.C, Hill, D.J, and Mayne, D.Q, "Design Issues in Adaptive Control," *IEEE Transactions on Automatic Control*, Vol. 33, No. 1, 1988, pp. 50–58.

[3] Anderson, B., and Dehghani, A., "Challenges of Adaptive Control: Past, Permanent, and Future," *Annual Reviews in Control*, Vol. 32, No. 2, 2008, pp. 123–135.

[4] Liu, Y., "Robust Adaptive Observer for Nonlinear Systems with Unmodeled Dynamics," *Automatica*, Vol. 45, No. 8, 2009, pp. 1891–1895.

[5] Hale, J., and Lunel, S. M. V., *Introduction to Functional Differential Equations*, Springer-Verlag, New York, 1993.

[6] Ioannou, P., and Kokotovic, P., "Instability Analysis and Improvement of Robustness of Adaptive Control," *Automatica*, Vol. 20, No. 5, 1984, pp. 583–594.

[7] Narendra, K., and Annaswamy, A., "A New Adaptive Law for Robust Adaptation Without Persistent Excitation," *IEEE Transactions on Automatic Control*, Vol. 32, No. 2, 1987, pp. 134–145.

[8] Bernstein, D.S, *Matrix Mathematics: Theory, Facts, and Formulas*, Princeton Univ. Press, Princeton, NJ, 2009, Chap. 9.

[9] Yucelen, T., and Calise, A.J, "Robustness of a Derivative-Free Adaptive Control Law," *AIAA Journal of Guidance, Control, and Dynamics*, Vol. 37, No. 5, Sept.-Oct. 2014, pp. 1583–1594.

[10] Yucelen, T., and Calise, A.J, "Derivative-Free Model Reference Adaptive Control," *AIAA Journal of Guidance, Control, and Dynamics*, Vol. 34, No. 4, July–Aug. 2011, pp. 933–950.

[11] Pomet, J.B, and Praly, L., "Adaptive Nonlinear Regulation: Estimation from the Lyapunov Equation," *IEEE Transactions on Automatic Control*, Vol. 37, No. 6, 1992, pp. 729–740.

[12] Xiao, B., Hu, Q., and Zhang, Y., "Fault-Tolerant Attitude Control for Flexible Spacecraft Without Angular Velocity Magnitude Measurement," *AIAA Journal of Guidance, Control, and Dynamics*, Vol. 34, No. 5, 2011, pp. 1556–1561.

CHAPTER 5

The Output Feedback Case

5.1 INTRODUCTION

Adaptive controllers can be classified as either full state feedback or output feedback. Output feedback adaptive controllers are required for applications in which it is impractical or impossible to sense the entire state of the process, such as active noise suppression, active control of flexible structures, fluid flow control systems, combustion control processes, and low cost or expendable unmanned aerial vehicle applications. Models for these applications vary from reasonably accurate low-frequency models, in the case of structural control problems, to less accurate low-order models in the case of active control of noise, vibrations, flows, and combustion processes.

There have been a number of results in the recent decades focused on output feedback adaptive controllers. For example, Esfandiari and Khalil [1] introduced a high gain observer for the reconstruction of the unavailable states. Krstic, Kanellakopoulos, and Kokotovic [2], and Marino and Tomei [3] proposed backstepping approaches for output feedback adaptive control, which are affine with respect to unknown parameters. In [4], Kim and Lewis suggested a neural network architecture for the observer. Output feedback adaptive control using a high gain observer and neural networks has also been proposed in [5] by Seshagiri and Khalil for nonlinear systems represented by input output models. Another method involving the design of an adaptive observer using function approximators and backstepping was given by Choi and Farrell [6] for a limited class of systems transformable to output feedback form in which nonlinearities depend on measurements only.

Calise, Hovakimyan, and Idan [7] developed an output feedback adaptive control architecture for nonlinear uncertain dynamical systems without using a state observer but with a stable low-pass filter to satisfy a strictly positive real (SPR) condition. It requires the input vector to the neural network be composed of current and past input-output data. In [8] Hovakimyan, Nardi, Calise, and Kim considered an output feedback adaptive controller with an error observer instead of a state observer. In this approach, only the relative degree of the regulated system needs to be known, and the basis functions employed depend only on the available input-output history. Volyanskyy, Haddad, and Calise [9] introduced Q-modification to output feedback adaptive control. This approach assumes a restrictive matching assumption.

Applying an adaptive control technique for uncertain dynamical systems in many cases implies replacement of an existing control system. On the other hand, it is highly desirable to consider an adaptive approach implementable in a form that augments an existing fixed gain or gain scheduled controller. This rationale has been one of the viewpoints taken in applying adaptive control in [10–14], on the basis it enhances the likelihood of certification, particularly in manned applications.

Hoagg, Santillo, and Bernstein [15] estimate Markov parameters leading to output feedback adaptive control algorithms with varying sets of assumptions. Lavretsky [16] introduced an adaptive output feedback design using asymptotic properties of linear quadratic Gaussian/loop transfer recovery (LQG/LTR) controllers that asymptotically satisfy an SPR condition. Our approach is similar in spirit to that of [16], with the exception that it is less restrictive because we do not attempt to satisfy an SPR condition, which naturally leads to a formulation relying on the existence of a positive-definite solution of a parameter-dependent Riccati equation. Therefore, our approach is also applicable to nonminimum phase systems.

This chapter extends the DF-MRAC development in Chapter 3 to the output feedback case using a state observer. The observer, which may be viewed as a part of the nominal control design, is employed in the adaptive part of the design in place of a reference model. Although a reference model is defined from a theoretical perspective, it need not be realized. A derivative-free weight update law in output feedback form ensures the estimated states follow both the reference model states and the true states, so both the state estimation error and the state tracking error are bounded. The approach is particularly advantageous for applications to systems that can undergo a sudden change in dynamics, such as might be due to reconfiguration, deployment of a payload, docking, structural damage, or a difficult-to-model disturbance.

The stability analysis uses a Lyapunov–Krasovskii functional entailing the solution of a parameter-dependent Riccati equation [17], rather than a Lyapunov equation, to show all the error signals are uniformly ultimately bound (UUB). It shows how the Riccati equation parameter influences the ultimate bounds for the state estimate error and the adapted weight error. The complexity of the approach is significantly less than many other approaches to output feedback adaptive control because it can be implemented in a form augmenting an existing observer-based linear control design. The design procedure is illustrated on a model of wing-rock dynamics.

Section 5.2 presents the observer-based derivative-free output feedback adaptive control architecture. Section 5.3 discusses the properties of a parameter-dependent Riccati equation. Section 5.4 provides the stability analysis for the architecture proposed in Section 5.2, based on the parameter-dependent Riccati equation described in Section 5.3. The error signals are shown to be UUB. Section 5.5 illustrates the proposed adaptive control approach on a simple second-order model. Section 5.6 introduces a method for reducing the effect of sensor noise, applicable to both the state feedback case and the output feedback case, and Section 5.7 points out the possible use of a linear version of DF-MRAC. Section 5.8 illustrates the use of adaptive loop transfer recovery on the simple example of Section 5.6. A summary is provided in Section 5.9.

5.2 CONTROL ARCHITECTURE

Consider the uncertain, nonlinear, and time-varying dynamical system given by

$$\dot{x}(t) = Ax(t) + B[u(t) + \Delta(t, x(t))] \tag{5.1}$$

$$y(t) = Cx(t) \tag{5.2}$$

where $A \in R^{nxn}$, $B \in R^{nxm}$, and $C \in R^{mxn}$ are known matrices; $x(t) \in R^n$ is the unknown state vector; $u(t) \in R^m$ is the control input vector; $\Delta : R\ x\ R^n \rightarrow R^m$ is the matched uncertainty; and $y(t) \in R^m$ is the regulated output vector. The pair (*A*, *B*) is controllable, the triple (*A*, *B*, *C*) is minimal, and the control input vector is restricted to the class of admissible controls consisting of measurable functions. Furthermore, the system has a sensed output vector denoted by

$$y_s(t) = C_s x(t) + D_s u(t) \tag{5.3}$$

where $y_s(t) \in R^l$, $C_s(t) \in R^{lxn}$, $l \geq m$, and the elements of $y(t)$ are a subset of the elements of $y_s(t)$. Only the sensed output vector is available for feedback.

Remark 5.1: The system given by Eqs. (5.1) and (5.2) assumes the control input vector and the regulated output vector have the same dimension. For the case when the dimension of the control input vector is larger than the dimension of the regulated output vector, and there is redundancy in actuation, one can use either matrix inverse or pseudoinverse approaches, constrained control allocation, pseudocontrols, or daisy chaining [18–20] to reduce the dimension of the control input vector to the dimension of the regulated output vector.

Remark 5.2: Although not presented here, the formulation can be extended to the case where there is uncertainty also in *B*, as was done in Section 3.4 in Chapter 3.

Consider the state observer–based nominal control law given by

$$u_n(t) = -K_1\hat{x}(t) + K_2 r(t) \tag{5.4}$$

where $K_1 \in R^{mxn}$ and $K_2 \in R^{mxm}$ are known feedback and feedforward matrices, respectively; $r(t) \in R^m$ is the regulated output command vector; and $\hat{x}(t) \in R^n$ is an observer estimate of $x(t)$ given by

$$\begin{aligned}\dot{\hat{x}}(t) &= A\hat{x}(t) + Bu_n(t) + L[y_s(t) - C_s\hat{x}(t) - D_su(t)] \\ &= (A - BK_1 - LC_s)\hat{x}(t) + L[y_s(t) - D_su(t)] + BK_2r(t) \end{aligned} \tag{5.5}$$

with $L \in R^{nxl}$ being the observer gain matrix, ensuring $A_e \triangleq A - LC_s \in R^{nxn}$ is Hurwitz. Define the reference model

$$\dot{x}_m(t) = A_m x_m(t) + B_m r(t) \tag{5.6}$$

where $x_m(t) \in R^n$, $A_m \triangleq A - BK_1$ is Hurwitz by design, and $B_m \triangleq BK_2$.

Remark 5.3: The above implies the gains K_1 and K_2 have been designed using the certainty equivalence principle ($\hat{x}(t) = x(t)$) with $\Delta(t, x(t)) = 0$, so that $C\hat{x}(t)$ tracks $r(t)$ to within some set of specifications on both the transient and steady state performance.

Remark 5.4: The reference model serves as an idealized measure of the desired closed loop performance and does not have to be realized as a part of the adaptive control system design.

Assumption 5.1: The matched uncertainty in Eq. (5.1) can be linearly parameterized as

$$\Delta(t, x(t)) = W^T(t)\beta(x(t)), \quad x \in D_x \subset R^n \tag{5.7}$$

where $W(t) \in R^{sxm}$ is an unknown time-varying weight matrix satisfying $\|W(t)\| < \overline{\omega}$, $\beta(x): R^n \rightarrow R^s$ is a known Lipschitz continuous basis vector of the form $\beta(x) = [\beta_1(x), \beta_2(x), \ldots, \beta_s(x)]^T$ with $|\beta(x)| \leq \overline{\beta}$, and D_x is sufficiently large.

The adaptive control objective is to design a control law $u(\cdot)$ for the system given by Eqs. (5.1) and (5.2) so that $x(t)$ tracks $x_m(t)$ with bounded error. For this purpose, the nominal control law $u_n(t)$ given by Eq. (5.4) is augmented with the adaptive control law $u_{ad}(t)$ as

$$u(t) = u_n(t) - u_{ad}(t) \tag{5.8}$$

$$u_{ad}(t) = \hat{W}^T(t)\beta(\hat{x}(t)) \tag{5.9}$$

The state observer given by Eq. (5.5) is regarded as a part of the nominal control design; however, our viewpoint is that L in Eq. (5.5) may be altered, if necessary, for purposes of adaptively augmenting the nominal controller.

Denote $\tilde{x}(t) \triangleq x(t) - \hat{x}(t)$ for the state estimation error, $\hat{e}(t) \triangleq \hat{x}(t) - x_m(t)$ for the estimated state tracking error, and $\tilde{W}(t) \triangleq W(t) - \hat{W}(t)$ for the weight estimate error. From Eqs. (5.1) and (5.5), the dynamics for the state estimation error can be written in the form

$$\dot{\tilde{x}}(t) = A_e\tilde{x}(t) + B\tilde{W}^T(t)\beta(\hat{x}(t)) + Bg(x(t), \hat{x}(t)) \tag{5.10}$$

where $g(x(t), \hat{x}(t)) \triangleq W^T(t)[\beta(\hat{x}(t)) - \beta(x(t))]$ with $|g(x(t), \hat{x}(t)| \leq \overline{w}L_\beta|\tilde{x}(t)|$, and $L_\beta > 0$ is the Lipschitz constant for the known basis vector. Likewise, from

Eqs. (5.5) and (5.6), the dynamics for the estimated state tracking error can be written in the form

$$\dot{\hat{e}}(t) = A_m \hat{e}(t) + LC\tilde{x}(t) \tag{5.11}$$

Next, consider the derivative-free weight update law given by

$$\hat{W}(t) = \Omega_1 \hat{W}(t-\tau) + \hat{\Omega}_2(t) \tag{5.12}$$

where $\tau > 0$. The matrices $\Omega_1 \in R^{sxs}$ and $\hat{\Omega}_2 : R^n x R^m \to R^{sxm}$ satisfy

$$0 \leq (1+\mu)\Omega_1^T \Omega_1 < I_s,\ \mu > 0 \tag{5.13}$$

$$\hat{\Omega}_2(t) \triangleq \kappa_2 \beta(\hat{x}(t))\tilde{y}^T(t), \quad \kappa_2 > 0 \tag{5.14}$$

where $\tilde{y}(t) \triangleq y(t) - \hat{y}(t)$ and $\hat{y}(t) = C\hat{x}(t)$. The condition in Eq. (5.13) is equivalent to $\underline{\lambda}\left(I - (1+\mu)\Omega_1^T\Omega_1\right) > 0$. Using Eq. (3.31) from Chapter 3, this can equivalently be written as $1 - (1+\mu)\overline{\sigma}^2(\Omega_1) > 0$, or $\overline{\sigma}(\Omega_1) < \sqrt{1/(1+\mu)}$.

A visualization of the DF-MRAC output feedback adaptive control architecture is given in Fig. 5.1. Note that the state observer takes on the role of the reference model in Fig. 3.1 in.Chapter 3. Its dynamics are the same as the reference model in Eq. (5.6) if $u_{ad}(t)$ cancels $\Delta(t, x(t))$, and in this case the observer error transient $\tilde{y}_s(t) \triangleq y_s(t) - C_s\hat{x}(t)$ goes to zero. Thus, the components added to an observer-based nominal controller architecture, in order to realize the adaptive control, reduces to implementing the adaptation law in Eqs. (5.12) and (5.14).

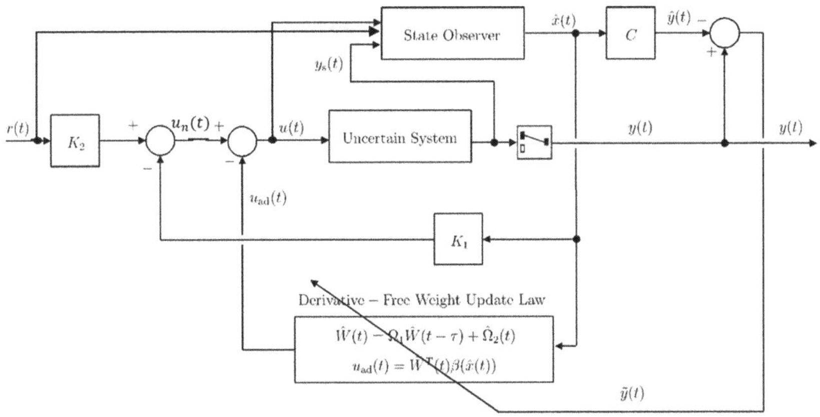

Fig. 5.1 DF-MRAC output feedback adaptive control architecture.

5.3 A PARAMETER-DEPENDENT RICCATI EQUATION

In this section we summarize the properties of a parameter-dependent Riccati equation [17] needed for the stability analysis in Section 5.4. Consider the following quadratic equation in P:

$$0 = A_e^T P + PA_e + Q \tag{5.15}$$

$$Q = Q_o + \nu NN^T,\ Q_o > 0 \tag{5.16}$$

$$N = C^T - PB \tag{5.17}$$

in which $Q_o \in R^{nxn}$ and $A_e = A - LC_s$.

Remark 5.5: Let $0 \leq \nu < \overline{\nu}$ define the largest set within which there exists a positive definite solution for P. Because $P > 0$ for $\nu = 0$ and P depends continuously on ν, the existence of $P(\nu) > 0$ for $0 \leq \nu < \overline{\nu}$ is assured.

Remark 5.6: If $N = 0$ in Eq. (5.17), then it follows from Eqs. (5.15–5.17) that

$$0 = A_e^T P_o + P_o A_e + Q_o \tag{5.18}$$

$$0 = C^T - P_o B \tag{5.19}$$

which implies the transfer function associated with the system $G(s) = C(sI - A_e)^{-1}B$ is positive real [21]. In this case, Eq. (5.15) reduces to a Lyapunov equation associated with the error dynamics in Eq. (5.10), which is commonly employed in the stability analysis of adaptive systems, and $\overline{\nu} = \infty$. This suggests that for the purposes of adaptive control design, when $m > 1$, it may be advantageous to define a new measurement by taking a linear combination of the existing measurements

$$y_0(t) = My(t) = MCx(t) = C_0 x(t) \tag{5.20}$$

where $M \in R^{mxm}$ is a norm-preserving transformation minimizing a norm measure of N_o where

$$N_o \triangleq C_0^T - P_o B \tag{5.21}$$

with P_o defined as the value of P satisfying Eq. (5.18). Taking the Frobenius norm as a measure, it can be shown that the solution for M minimizing $||N_o||$ subject to the constraint $||MC|| = ||C||$ is given by

$$M = kM_0 \tag{5.22}$$

$$M_0^T = [CC^T]^{-1} CP_o B \tag{5.23}$$

$$k = ||C|| / ||M_0 C|| \tag{5.24}$$

Lemma 5.1: If A_m has no repeated eigenvalues and the state observer gain matrix L of Eq. (5.5) is designed using pole placement such that

$$\lambda(A_e) = k\lambda(A_m) \tag{5.25}$$

then for $\nu < \overline{\nu}$, we can denote $P(k)$ as the corresponding positive definite solution for P of the parameter-dependent Riccati equation given by Eqs. (5.15–5.17). In this case,

$$\lim_{k\to\infty} P(k) = 0 \tag{5.26}$$

Proof: Let S_m and S_e be diagonalizing transformations for A_m and A_e, respectively. Then

$$A_m = S_m \text{diag}[\lambda(A_m)] S_m^{-1} \tag{5.27}$$

$$A_e = kS_e \text{diag}[\lambda(A_e)] S_e^{-1} \tag{5.28}$$

With $S_s \triangleq S_e S_m^{-1}$, A_e can be written as

$$A_e = kS_s A_m S_s^{-1} = kA_s \tag{5.29}$$

Using Eq. (5.28) and denoting $P_k \triangleq kP(k)$ and $N_k \triangleq C^T - k^{-1}P_k B$, the parameter-dependent Riccati equation in Eqs. (5.15–5.17) can be written as

$$0 = A_s^T P_k + P_k A_s + Q_0 + \nu N_k N_k^T \tag{5.30}$$

Taking the limit of Eq. (5.30) as $k \longrightarrow \infty$, then

$$0 = A_s^T P_\infty + P_\infty A_s + Q_0 + \nu CC^T \tag{5.31}$$

The solution of Eq. (5.31) is finite, so for all finite k, $P(k) > 0$, and

$$\lim_{k\to\infty} P(k) = \lim_{k\to\infty} P_k/k \tag{5.32}$$

Remark 5.7: Lemma 5.1 implies $\|PB\|$ can be made arbitrarily small by making the observer dynamics sufficiently fast, while preserving the property $P > 0$. This property is of importance because it reduces the ultimate bound, as noted later in Remark 5.8.

The next lemma shows that for $\nu < \overline{\nu}$, Eqs. (5.15–5.17) can reliably be solved for $P > 0$ using the Potter approach given in [22]. This also implies $\overline{\nu}$ can be determined by searching for the boundary value resulting in a failure of the algorithm to converge. In stating the lemma, we employ the notation ric$(\cdot)$ and dom$(\cdot)$, as defined in [23].

Lemma 5.2: Let P satisfy the parameter-dependent Riccati equation given by Eqs. (5.15–5.17), and let the modified Hamiltonian be given by

$$H \triangleq \begin{bmatrix} A_e - \nu BC & \nu R \\ -Q & -(A_e - \nu BC)^T \end{bmatrix} \tag{5.33}$$

where $Q \triangleq Q_0 + \nu CC^T$ and $R \triangleq BB^T$. Then, for all $0 < \nu < \overline{\nu}$, $H \in \text{dom(ric)}$ and $P = \text{ric}(H)$.

Proof: The proof follows from Lemmas 1 and 2 in [23].

5.4 STABILITY ANALYSIS

This section presents the main theorem concerning DF-MRAC for the output feedback architecture described in Section 5.2, and includes a related lemma and corollary.

Theorem 5.1: Consider the controlled uncertain dynamical system given by Eqs. (5.1) and (5.2), subject to Assumption 5.1. Consider, in addition, the feedback control law given by Eq. (5.8), with the nominal feedback control component given by Eqs. (5.4) and (5.5), and the adaptive feedback control component given by Eqs. (5.9) and (5.12), subject to the conditions in Eqs. (5.13) and (5.14), and with $\kappa_2 < \overline{\kappa}_2 \triangleq \overline{\nu}/\overline{\beta}^2 - 1/\mu$. Then, $\tilde{x}(t)$ and $\tilde{W}(t)$ are UUB.

Proof: Using Eq. (5.12) and defining, as before in Eq. (3.5),

$$\Omega_2(t) \triangleq W(t) - \Omega_1 W(t-\tau) \tag{5.34}$$

where $||\Omega_2(t)|| \leq \delta^*$, $\delta^* \triangleq w^*(1 + ||\Omega_1||)$, the weight update error can be rewritten as

$$\tilde{W}(t) = \Omega_1 \tilde{W}(t-\tau) + \Omega_2(t) - \hat{\Omega}_2(t) \tag{5.35}$$

Using Eq. (3.35), the state estimation error in Eq. (5.10) becomes

$$\dot{\tilde{x}}(t) = A_e\tilde{x}(t) + B[\Omega_1\tilde{W}(t-\tau) + \Omega_2(t) - \hat{\Omega}_2(t)]^T\beta(\hat{x}(t)) + Bg(x(t), \hat{x}(t)) \tag{5.36}$$

$\hat{\Omega}_2(t)$ in Eq. (5.14) can be equivalently written using Eq. (5.17) as

$$\hat{\Omega}_2(t) = \kappa_2\beta(\hat{x}(t))\tilde{x}^T(t)BC^T = \kappa_2\beta(\hat{x}(t))\tilde{x}^T(t)[PB + N] \tag{5.37}$$

For $N = 0$, this is similar in form to Eq. (3.4) in Chapter 3, in the case of state feedback.

To show the closed-loop system given by Eqs. (5.35) and (5.36) is UUB, consider the Lyapunov–Krasovskii functional

$$L_k(e(t), \tilde{W}_t) = \tilde{x}^T P\tilde{x}(t) + \rho\, tr\left[\int_{t-\tau}^{t} \tilde{W}^T(s)\tilde{W}(s)ds\right], \quad \rho = 1/\kappa_2 \tag{5.38}$$

which is identical in form to Eq. (3.8) in Chapter 3, except $\tilde{x}(t)$ is used in place of $e(t)$. The directional derivative of Eq. (5.38) along the closed-loop system trajectories of Eqs. (5.35) and (5.36) is given by

$$\begin{aligned}\dot{L}_k(e(t), \tilde{W}_t) = &-\tilde{x}^T(t)Q\tilde{x}(t) + 2\tilde{x}^T(t)PBg(x(t), \hat{x}(t)) + 2\tilde{x}^T(t)PB[\Omega_1\tilde{W}(t-\tau)]^T\beta(\hat{x}(t)) \\ &- 2\tilde{x}^T(t)PB\hat{\Omega}_2^T(t)\beta(\hat{x}(t)) + 2\tilde{x}^T(t)PB\Omega_2^T(t)\beta(\hat{x}(t)) \\ &+ \rho tr[\tilde{W}^T(t)\tilde{W}(t) - \tilde{W}^T(t-\tau)\tilde{W}(t-\tau)]\end{aligned} \tag{5.39}$$

Note the second term in Eq. (5.39) does not appear in the corresponding expression in Eq. (3.9) because $g(x(t), \hat{x}(t)) = 0$ in the state feedback case. Using Eq. (5.35) to expand the term $\rho\, tr[\tilde{W}^T(t)\tilde{W}(t)]$ in

Eq. (5.39) produces

$$
\begin{aligned}
\dot{L}_k(e(t), \tilde{W}_t) = & -\tilde{x}^T(t)Q\tilde{x}(t) + 2\tilde{x}^T(t)PBg(x(t), \hat{x}(t)) + 2\tilde{x}^T(t)PB[\Omega_1\tilde{W}(t-\tau)]^T\beta(\hat{x}(t)) \\
& - 2\tilde{x}^T(t)PB\hat{\Omega}_2^T(t)\beta(\hat{x}(t)) + 2\tilde{x}^T(t)PB\Omega_2^T(t)\beta(\hat{x}(t)) \\
& \rho\, tr[\tilde{W}^T(t-\tau)\Omega_1^T\Omega_1\tilde{W}(t-\tau) + \hat{\Omega}_2^T(t)\hat{\Omega}_2(t) + \Omega_2^T(t)\Omega_2(t) \\
& - 2\hat{\Omega}_2^T(t)\Omega_1\tilde{W}(t-\tau) + 2\tilde{W}^T(t-\tau)\Omega_1^T\Omega_2(t) \\
& - 2\hat{\Omega}_2^T(t)\Omega_2(t) - \tilde{W}^T(t-\tau)\tilde{W}(t-\tau)]
\end{aligned}
\tag{5.40}
$$

The next analysis step is identical to the steps leading from Eq. (3.10) to Eq. (3.12) in Chapter 3, except for the five additional terms involving N when using Eq. (5.37) to substitute for $\hat{\Omega}_2(t)$ in the fourth, seventh, ninth, and 11th terms in Eq. (5.40). However, dependency on N in the fourth term cancels with the linear-in-N portion of the seventh term, leaving only three terms involving N. As a result of these algebraic steps, we have

$$
\begin{aligned}
\dot{L}_k(e(t), \tilde{W}_t) \le & -\tilde{x}^T(t)Q\tilde{x}(t) + 2\tilde{x}^T(t)PBg(x(t), \hat{x}(t)) \\
& - \rho tr\left[\tilde{W}^T(t-\tau)\left\{I_s - (1+\gamma)\Omega_1^T\Omega_1\right\}\tilde{W}(t-\tau)\right] \\
& + \rho\left(1 + \frac{1}{\gamma}\right) tr[\Omega_2^T(t)\Omega_2(t)] \\
& - tr\left[2N^T\tilde{x}(t)\beta^T(\hat{x}(t))\Omega_2\right] + \kappa_2\tilde{x}(t)^T NN^T\tilde{x}(t)\overline{\beta}^2 \\
& - 2tr[N^T\tilde{x}(t)\beta^T(\hat{x}(t))\Omega_1\tilde{W}(t-\tau)]
\end{aligned}
\tag{5.41}
$$

where the parameter $\gamma > 0$ is associated with the first application of Young's inequality in Chapter 3, Eq. (3.11). Applying Young's inequality again for the last term in Eq. (5.41) produces

$$
\begin{aligned}
2tr[N^T\tilde{x}(t)\beta^T(\hat{x}(t))\Omega_1\tilde{W}(t-\tau)] \le & tr\left[\mu\tilde{W}^T(t-\tau)\Omega_1{}^T\Omega_1\tilde{W}(t-\tau)\right] \\
& + \tilde{x}(t)^T NN^T\tilde{x}(t)/\mu, \quad \mu > \overline{\beta}^2/\,\overline{\nu}
\end{aligned}
\tag{5.42}
$$

The condition $\mu > \overline{\beta}^2/\,\overline{\nu}$ is imposed to ensure $\overline{\kappa}_2 > 0$. Using Eq. (5.42) in Eq. (5.41) results in

$$
\begin{aligned}
\dot{L}_k(e(t), \tilde{W}_t) \le & -\tilde{x}^T(t)Q\tilde{x}(t) + 2\tilde{x}^T(t)PBg(x(t), \hat{x}(t)) \\
& - \rho\, tr\left[\tilde{W}^T(t-\tau)\left\{I_s - (1+\mu+\gamma)\Omega_1^T\Omega_1\right\}\tilde{W}(t-\tau)\right] \\
& + \rho\left(1 + \frac{1}{\gamma}\right) tr\left[\Omega_2^T(t)\Omega_2(t)\right] \\
& - tr\left[2N^T\tilde{x}(t)\beta^T(\hat{x}(t))\Omega_2\right] + (\kappa_2 + 1/\mu)\overline{\beta}^2\tilde{x}(t)^T NN^T\tilde{x}(t)
\end{aligned}
\tag{5.43}
$$

Applying the definitions of Q and ν in Eq. (5.16), with $\nu = (\kappa_2 + 1/\mu)\overline{\beta}^2$, the first and last terms in Eq. (5.41) can be combined to produce

$$\begin{aligned} \dot{L}_k(e(t), \tilde{W}_t) \leq & -\tilde{x}^T(t)Q_o\tilde{x}(t) + 2\tilde{x}^T(t)PBg(x(t), \hat{x}(t)) \\ & - \rho\, tr\left[\tilde{W}^T(t-\tau)\left\{I_s - (1+\mu+\gamma)\Omega_1^T\Omega_1\right\}\tilde{W}(t-\tau)\right] \\ & + \rho(1+1/\gamma)tr\left[\Omega_2^T(t)\Omega_2(t)\right] - tr[2N^T\tilde{x}(t)\beta^T(\hat{x}(t))\Omega_2] \end{aligned} \quad (5.44)$$

where the condition $\kappa_2 < \overline{\kappa}_2$ ensures $\nu < \overline{\nu}$. Using $|g(x(t), \hat{x}(t))| \leq \overline{g} \equiv 2\overline{\omega}\overline{\beta}$ yields

$$\dot{L}_k(e(t), \tilde{W}_t) \leq -c_1|\tilde{x}(t)|^2 - c_2\left\|\tilde{W}(t)^2\right\| - c_3\left\|\tilde{W}(t-\tau)^2\right\| + d_1 + d_2|\tilde{x}(t)| \quad (5.45)$$

where

$$c_1 = \underline{\lambda}(Q_o) > 0 \quad (5.46)$$
$$c_2 = 0 \quad (5.47)$$
$$c_3 = \underline{\lambda}(I - (1+\mu+\gamma)\Omega_1^T\Omega_1 > 0,\ 0 < \gamma < \overline{\gamma} \quad (5.48)$$
$$d_1 = \rho(1+1/\gamma)\overline{\delta}^2 \quad (5.49)$$
$$d_2 = 2(\|PB\|\overline{g} + \|N\|\|\Omega_2\|\overline{\beta}) \quad (5.50)$$

The existence of $\overline{\gamma} > 0$ in Eq. (5.48) follows from the condition in Eq. (5.13), ensuring the inequality is satisfied for $\gamma = 0$. We can further arrange Eq. (5.45) as

$$\dot{L}_k(e(t), \tilde{W}_t) \leq = -[\sqrt{c_1}|\tilde{x}(t)| - d_2/2\sqrt{c_1}]^2 - c_3\tilde{W}(t-\tau)^2 + d \quad (5.51)$$

where $d = d_1 + d_2{}^2/4c_1$. This reveals that $|\tilde{x}(t)| > \Psi_1$ or $\left\|\tilde{W}(t-\tau)\right\| > \Psi_3$ renders $\dot{L}_k(e(t), \tilde{W}_t) < 0$, where $\Psi_1 = \sqrt{d/c_1} + d_2/2c_1$ and $\Psi_3 = \sqrt{d/c_3}$, and therefore both $\tilde{x}(t)$ and $\tilde{W}(t)$ are UUB. ■

Remark 5.8: Under the conditions of Theorem 5.1, Corollary 3.1 also applies to the output feedback problem. Note that reducing $\|PB\|$ as mentioned in Remark 5.7 reduces the magnitude of d_2 and therefore the magnitude of both Ψ_1 and Ψ_3 in Chapter 3, Eq. (3.20). Extensions of Theorems 3.2, 3.3, and 3.4 in Chapter 3 to the output feedback problem are recommended among the future research topics in Chapter 8.

Lemma 5.3: If $\tilde{x}(t)$ is bounded, then the state tracking error $e(t) \triangleq x(t) - x_m(t)$ is bounded.

Proof:

$$\begin{aligned} |e(t)| &= |x(t) - x_m(t)| \\ &= |x(t) - \hat{x}(t) + \hat{x}(t) - x_m(t)| \\ &\leq |x(t) - \hat{x}(t)| + |\hat{x}(t) - x_m(t)| \\ &= |\tilde{x}(t)| + |\hat{e}(t)| \end{aligned} \quad (5.52)$$

where if $\tilde{x}(t)$ is bounded, then from Eq. (5.11), $\hat{e}(t)$ is also bounded. This implies that $e(t)$ is also bounded. ■

Corollary 5.1: Consider the system of equation given by Eqs. (5.10) and (5.11). If $\tilde{x}(t)$ is UUB by r, then $e(t)$ is UUB by $r(1+\nu)$ where

$$\nu \triangleq 2\|P_m LC\|/\underline{\lambda}(Q_m) \tag{5.53}$$

and $P_m \in R^{nxn}$ being the positive-definite solution to the Lyapunov equation given by

$$0 = A_m^T P_m + P_m A_m + Q_m \tag{5.54}$$

for any given positive-definite matrix $Q_m \in R^{nxn}$.

Proof: To show $\hat{e}(t)$ is UUB, consider the Lyapunov function given by

$$L(\hat{e}(t)) = \hat{e}(t)^T P_m \hat{e}(t) \tag{5.55}$$

where P_m satisfies Eq. (5.54). We can write the directional derivative of Eq. (5.55) along the trajectories of $\hat{e}(t)$ given by Eq. (5.11) as

$$\begin{aligned} \dot{L}(\hat{e}(t)) &= -\hat{e}^T(t)Q_m\hat{e}(t) + 2\hat{e}^T(t)P_m LC\tilde{x}(t) \\ &\le -|\hat{e}(t)|[\underline{\lambda}(Q_m)|\hat{e}(t)| - 2\|P_m LC\||\tilde{x}(t)|] \end{aligned} \tag{5.56}$$

so $\dot{L}(\hat{e}(t)) \le 0$ for all $|\hat{e}(t)| > r\nu$, because $\tilde{x}(t)$ is UUB by r and ν is given by Eq. (5.53). Hence, it follows from Lemma 5.3 that $|e(t)|$ is UUB by $r(1+\nu)$. ■

5.5 A SIMPLE EXAMPLE

This section presents an example of the architecture described in Fig. 5.1 using a simple second-order model for an uncertain plant model in the form

$$\begin{bmatrix} \dot{x}_1(t) \\ \dot{x}_2(t) \end{bmatrix} = \begin{bmatrix} 0 & 1 \\ 0 & 0 \end{bmatrix} \begin{bmatrix} x_1(t) \\ x_2(t) \end{bmatrix} + \begin{bmatrix} 0 \\ 1 \end{bmatrix} [u(t) + \Delta(t, x(t))] \tag{5.57}$$

$$y(t) = y_s(t) = [1 \quad 0] \begin{bmatrix} x_1(t) \\ x_2(t) \end{bmatrix} + w(t) \tag{5.58}$$

$$\begin{aligned} \Delta(t, x(t)) = {} & [\alpha_1 + f_1(t)]x_1(t) + [\alpha_2 + f_2(t)]x_2(t) + \alpha_3|x_1(t)|x_2(t) \\ & + \alpha_4|x_2(t)|x_2(t) + \alpha_5 x_1^3(t) + d(t) \end{aligned} \tag{5.59}$$

with constant coefficients $\alpha_1 = 0.25$, $\alpha_2 = 0.7$, $\alpha_3 = -0.6$, $\alpha_4 = 0.01$, and $\alpha_5 = 0.02$; time-varying coefficients $f_1(t) =$ a square wave with amplitude $\pm$ 0.5 and 15-second period; $f_2(t) = 0.5\sin(t)$, a square wave external disturbance $d(t)$ with amplitude $\pm$ 1 and 10-second period; and sensor noise $w(t)$. This model can be viewed as representative of aircraft wing-rock dynamics [24]. The reference model is selected to be second order with a natural frequency of 1.6 rad/s and a damping ratio of 0.8, and to have a unity gain from $r(t)$ to $y_m(t)$ at low frequency. This amounts to choosing $K_1 = [2.56, 2.56]$ and $K_2 = 2.56$. We also chose $L^T = [12.8\ 64.0]$, which corresponds to $k = 5$ in Eq. (5.25).

For the adaptive control design, we used sigmoidal basis functions of the form $\beta(x(t)) = [1.0,\ \beta_1(x_{1n}(t)),\ \beta_2(x_{2n}(t))]^T$, where $x_{in}(t) = x_i(t)/5$ are normalized inputs and $\beta_i(x_{in}(t)) = (1 - e^{-3x_{in}(t)})/(1 + e^{-3x_{in}(t)})$, i = 1, 2. The inputs are normalized by their maximum expected range so that the basis functions are nearly saturated when $x_i(t) = 5$. $|\beta_i(x_i)| \leq 1$, so it follows that $\overline{\beta} = \text{sqrt}(3)$. For $Q_o = 0.25 I_2$, it was determined by Lemma 5.2 that $\overline{\nu} = 124.4$. The value of $\mu > \overline{\beta}^2 / \overline{\nu} = 0.024$ ensures $\overline{\kappa}_2 > 0$. It also influences the limit on Ω_1. Selecting

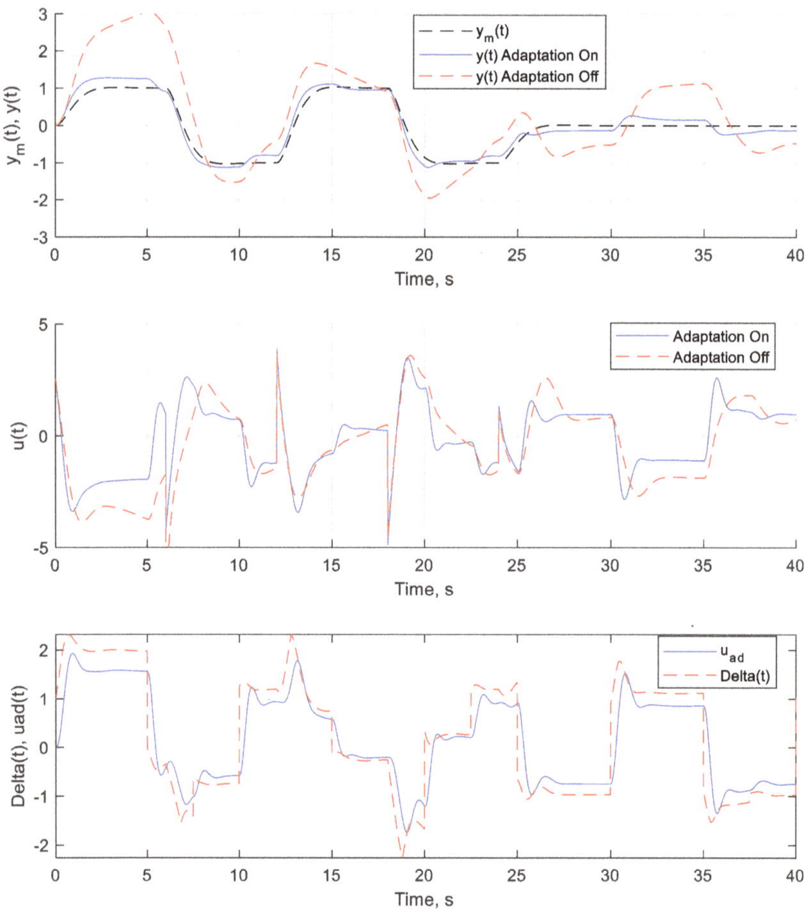

Fig. 5.2 Nominal and DF-MRAC responses in the absence of sensor noise.

$\mu = 0.05$ implies the adaptation gain in Eq. (5.14) must satisfy $\kappa_2 < \overline{\kappa}_2 = 21.5$, so we set κ_2 to half this amount as a margin of safety. Also, $\Omega_1 = -0.95I_3$, which, using the relation below Eq. (5.14), is chosen to satisfy Eq. (5.13), and we tuned the time delay parameter to $\tau = 0.05$ seconds. Both the controller and observer were implemented in discrete time using a step size of 0.01 seconds, and the plant model was simulated in essentially continuous time, with a step size of 0.001 seconds.

Consider the command tracking problem with initial conditions for the dynamics in Eq. (5.57) set to zero, and the command consisting of a series of step changes occurring every 7 seconds. Figure 5.2 compares the nominal (adaptation off) and DF-MRAC (adaptation on) responses in the absence of sensor

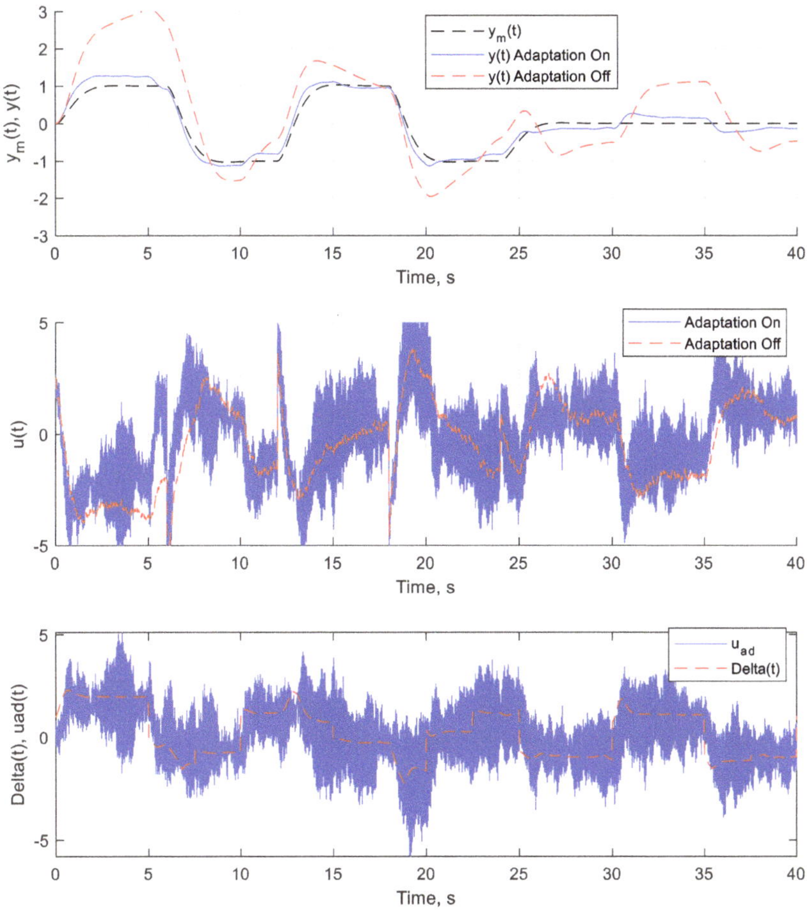

Fig. 5.3 Nominal and DF-MRAC responses in the presence of sensor noise.

noise. Overall performance is significantly improved with adaptation. Improvement is measured by comparing how well $y(t)$ follows $y_m(t)$, the response of the reference model, which under ideal conditions (no uncertainty) is followed with zero tracking error. Also shown at the bottom of Fig. 5.2 is a comparison between the $u_{ad}(t)$ and $\Delta(t, x(t))$.

Next we include the effect of sensor noise. To model sensor noise we let $w(t)$ be a bandlimited white noise process with a correlation time constant of 0.01 seconds and a noise power level of $1.0e^{-5}$. Figure 5.3 shows tracking performance is significantly improved with adaptation, but the effect of sensor noise is greatly amplified in comparison to the nominal control due to the fact that it has a direct pathway into $u_{ad}(t)$. This issue in adaptive control is addressed in the next section.

5.6 REDUCING THE EFFECT OF SENSOR NOISE

The DF-MRAC law with sensor noise reduction is nearly identical to Eqs. (5.12–5.14), except it uses a filtered version of the output error. This amounts to replacing Eq. (5.14) with

$$\hat{\Omega}_2(t) \triangleq \kappa_2 \beta(\hat{x}(t)) \tilde{y}_f^T(t), \quad \kappa_2 > 0 \tag{5.60}$$

where $\tilde{y}_f^T(t)$ is a filtered version of $\tilde{y}(t)$. In the case of a first-order filter, we have

$$\tau_f \dot{\tilde{y}}_f(t) = -\tilde{y}_f(t) + \tilde{y}(t) \tag{5.61}$$

where τ_f is the filter time constant. For simplicity of exposition we have assumed in Eq. (5.61) the same first-order filter is applied to each element of $\tilde{y}(t)$; however, the following arguments apply equally well when the order and/or the bandwidths of the filters applied to each element of $\tilde{y}(t)$ are different. We are concerned here with the properties of the adaptive controller for sufficiently large filter bandwidth (sufficiently small τ_f in the case of a first-order filter).

Theorem 5.2: Consider the controlled uncertain dynamical system given by Eqs. (5.1) and (5.2) subject to Assumption 5.1. Consider, in addition, the feedback control law given by Eq. (5.8), with the nominal feedback control component given by Eqs. (5.4) and (5.5), and the adaptive feedback control component given by Eqs. (5.9) and (5.12) subject to the conditions in Eqs. (5.13) and (5.14), and with $\kappa_2 < \overline{\kappa}_2$. Then there exists a finite $\overline{\tau}_f$ such that $\forall\ 0 < \tau_f < \overline{\tau}_f$, $\tilde{x}(t)$ and $\tilde{W}(t)$ are UUB.

Proof: The equations stated in Theorem 5.2 have the form of a singularly perturbed system with τ_f viewed as a small parameter. According to Tikhonov's theorem ([25], Theorem 11.1), an approximation for the solutions for $\tilde{x}(t, \tau_f)$ and $\tilde{W}(t, \tau_f)$ can be constructed from the solution of the reduced system obtained by setting $\tau_f = 0$ in Eq. (5.42), in which case $\tilde{y}_f(t) = \tilde{y}(t)$ and Theorem 5.1 applies to the reduced system. A requirement for the application of Tikhonov's

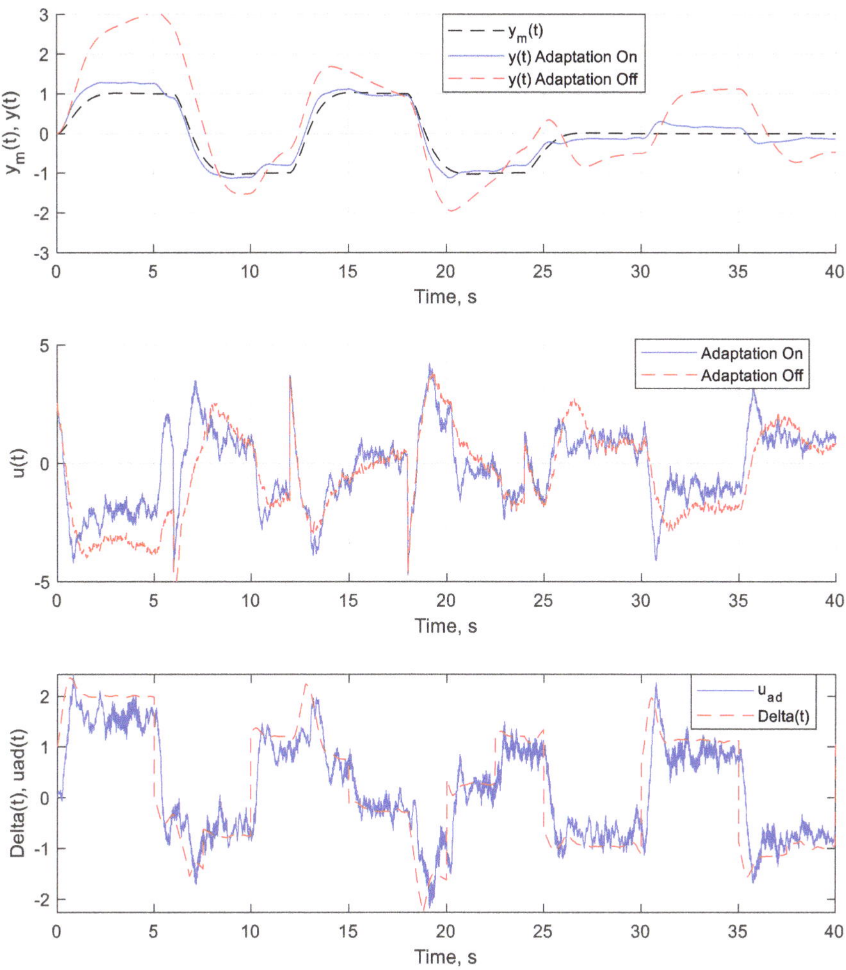

Fig. 5.4a Nominal and DF-MRAC responses using Eq. (5.60) in place of Eq. (5.14) with $\omega_{nf} = 40$ and $\zeta_{nf} = 0.8$.

theorem is that the equilibrium point of the boundary-layer system is exponentially stable uniformly in $\tilde{y}(t)\ \forall t \geq 0$. The boundary-layer system is obtained by applying the time transformation $t_{bl} = t/\tau_f$ to Eq. (5.61), which results in

$$\dot{\tilde{y}}_f(t_{bl}) = -\tilde{y}_f(t_{bl}) + \tilde{y}(t) \tag{5.62}$$

From Eq. (5.62) it is evident the exponential stability requirement is satisfied with $\tilde{y}(t)$ viewed as the instantaneous equilibrium point $\forall\ t \geq 0$. The main result of Tikhonov's theorem is there exists a finite

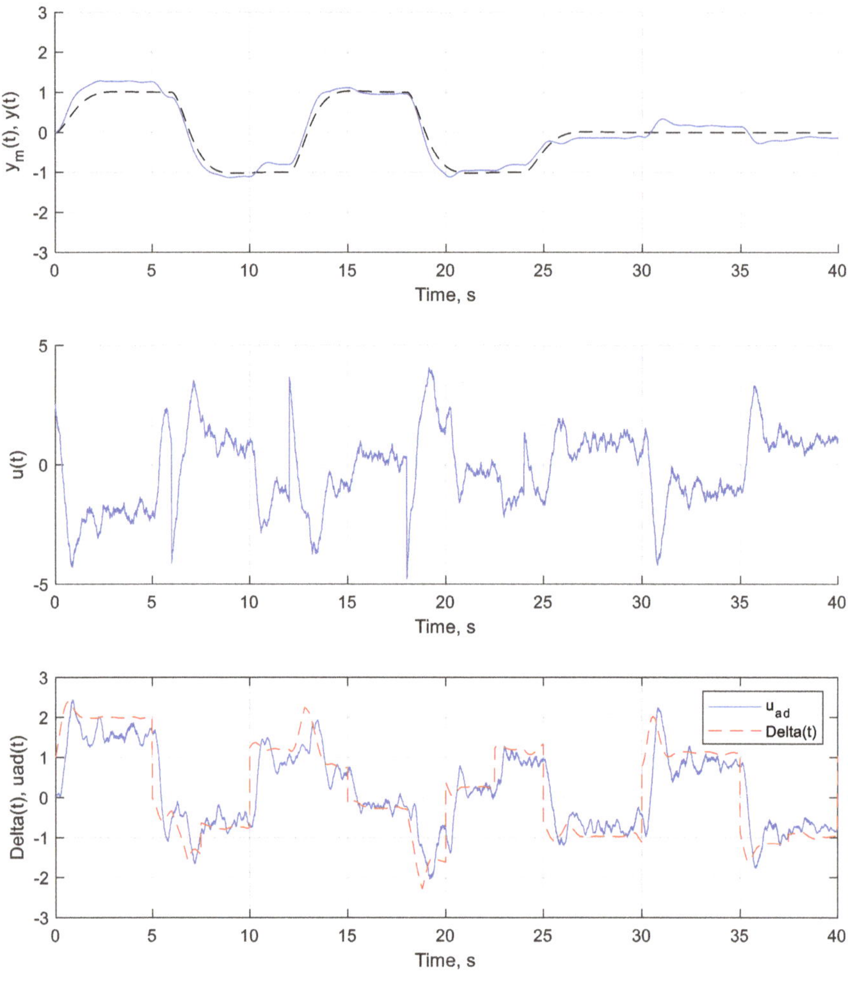

Fig. 5.4b Nominal and DF-MRAC responses using Eq. (5.60) in place of Eq. (5.14) with $\omega_{nf} = 20$ and $\zeta_{nf} = 0.8$.

$\bar{\tau}_f$ such that $\forall\ 0 < \tau_f < \bar{\tau}_f$

$$\tilde{x}(t, \tau_f) = \tilde{x}_r(t) + \mathcal{O}(\tau_f) \tag{5.63}$$

$$\tilde{W}(t, \tau_f) = \tilde{W}_r(t) + \mathcal{O}(\tau_f) \tag{5.64}$$

where $\mathcal{O}(\tau_f)$ signifies a term of the order of τ_f and subscript r denotes the reduced solution. Therefore, both $\tilde{x}(t, \tau_f)$ and $\tilde{W}(t, \tau_f)$ are UUB.

To illustrate the effectiveness of this approach to sensor noise reduction, we return to the example in Section 5.5 for the case with sensor noise. Figure 5.4a shows the response when $\tilde{y}_f^T(t)$ is used in place of $\tilde{y}^T(t)$, as explained in Eq. (5.60). In this case, a second-order filtering of $\tilde{y}(t)$ was used with a natural frequency of $\omega_{nf} = 40$ rad/s and a damping ratio of $\zeta_{nf} = 0.8$, which in continuous time satisfies

$$\ddot{\tilde{y}}_f(t) + 2\zeta_{nf}\omega_{nf}\dot{\tilde{y}}_f(t) + \omega_{nf}^2\tilde{y}_f(t) = \omega_{nf}^2\tilde{y}(t) \tag{5.65}$$

The corresponding filter in discrete time was implemented, again using a step size of 0.01 seconds. Note the dramatic improvement, in comparison to the responses for $u(t)$ and $u_{ad}(t)$ in Fig. 5.3, obtained without filtering of $\tilde{y}(t)$. An even smoother response is attainable by choosing $\omega_{nf} = 20$ rad/s. This is shown in Fig. 5.4b.

5.7 LINEAR DF-MRAC

One of the features unique to DF-MRAC that applies to all the formulations presented thus far is the fact Assumption 5.1 does not require the basis vector $\beta(x(t))$ in Eq. (5.7) to be state dependent. It can simply be chosen as a scalar constant, in which case the ideal weight vector becomes $W(t) \in R^{sx1}$. This is because in DF-MRAC we allow the ideal weights to be individually time varying. Letting $\beta(x(t)) = 1$ reduces the filtered adaptation law in Eq. (5.60) to

$$\hat{\Omega}_2(t) \triangleq \kappa_2\tilde{y}_f^T(t), \quad \kappa_2 > 0 \tag{5.66}$$

The adaptive control in Eq. (5.9) reduces to the following linear filtered form:

$$u_{ad}(t) = \hat{W}^T(t) = \Omega_1\hat{W}^T(t-\tau) + \kappa_2\tilde{y}_f(t) \tag{5.67}$$

Likewise, in the case of state feedback, the adaptive control reduces to

$$u_{ad}(t) = \hat{W}^T(t) = \Omega_1\hat{W}^T(t-\tau) + \kappa_2 B^T Pe(t) \tag{5.68}$$

which follows from Eqs. (3.2) and (3.4). This removes the burden of having to select a state-dependent parameterization of the uncertainty. It also doesn't matter how the bias term is selected because it simply multiplies the adaptation gain, κ_2; however, the choice of κ_2 must still satisfy the condition in Theorem 5.1.

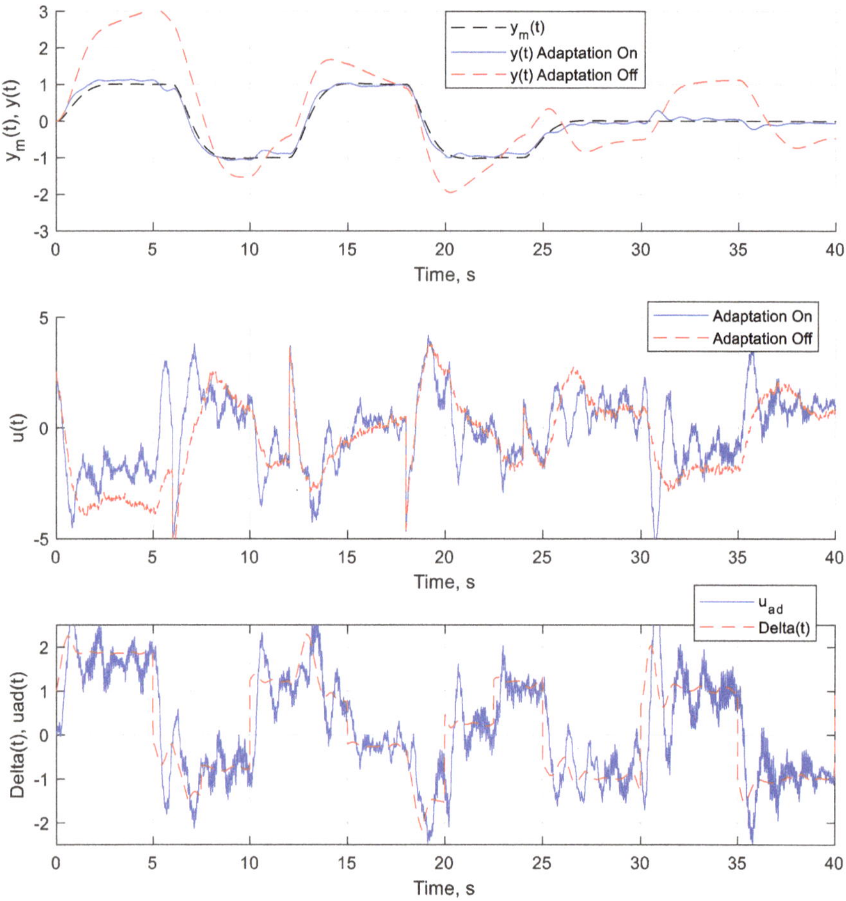

Fig. 5.5 Nominal and linear DF-MRAC responses using Eq. (5.66) in place of Eq. (5.60).

To illustrate the performance of linear DF-MRAC, we return once again to the simple example of Section 5.5 with $\beta(x(t)) = 1$. With this choice we find $\overline{\kappa}_2 = 104$. This time we set $\kappa_2 = \overline{\kappa}_2/4$ as a margin of safety and repeat the same example with filtering of $\tilde{y}(t)$ as explained in Section 5.6. In order to reduce the effect of sensor noise to approximately the same level it was necessary to reduce the filter natural frequency from $\omega_{nf} = 40$ rad/s to $\omega_{nf} = 20$ rad/s while maintaining the same damping ratio of $\zeta_{nf} = 0.8$. This reduction in bandwidth and the setting $\kappa_2 = \overline{\kappa}_2/4$ is to be expected because we are now relying entirely on the time variation in $\hat{W}^T(t)$ to cancel the effect of uncertainty and to follow reference commands. The resulting response is shown in Fig. 5.5,

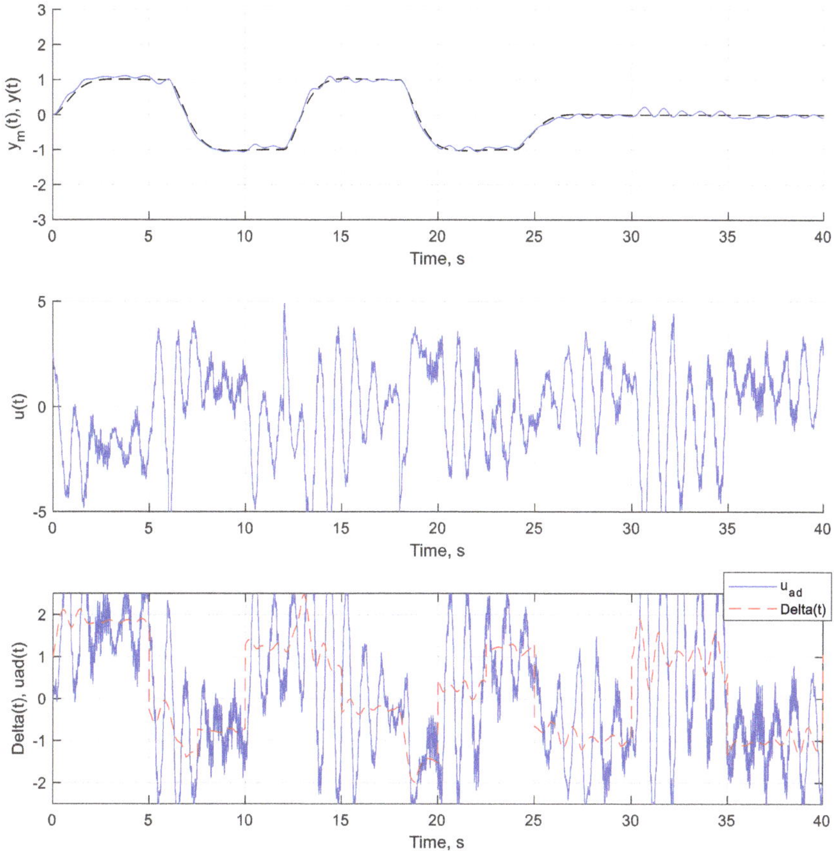

Fig. 5.6 DF-MRAC responses using $\kappa_2 = \overline{\kappa}_2/2$.

which compares favorably with the response shown in Fig. 5.4b that employed $\beta(x(t) = [1.0,\ \beta_1(x_{1n}(t)),\ \beta_2(x_{2n}(t))]^T$ as a basis function vector.

It is also of interest to examine what happens if the adaptation gain is increased; for example, if $\kappa_2 = \overline{\kappa}_2/2$ is used, the response in Fig. 5.6 results. Command tracking is improved slightly but at a large expense in control activity. The only way to improve tracking performance from the perspective of steady state error, without introducing excessive control activity, is to design a P+I form of nominal controller (see Chapter 7, Section 7.1); however, this may increase settling time and/or percent overshoot in response to a step command. If $\kappa_2 = 0.95\,\overline{\kappa}_2$ is used, the error response remains ultimately bounded but is now unacceptably large, as shown in Fig. 5.7.

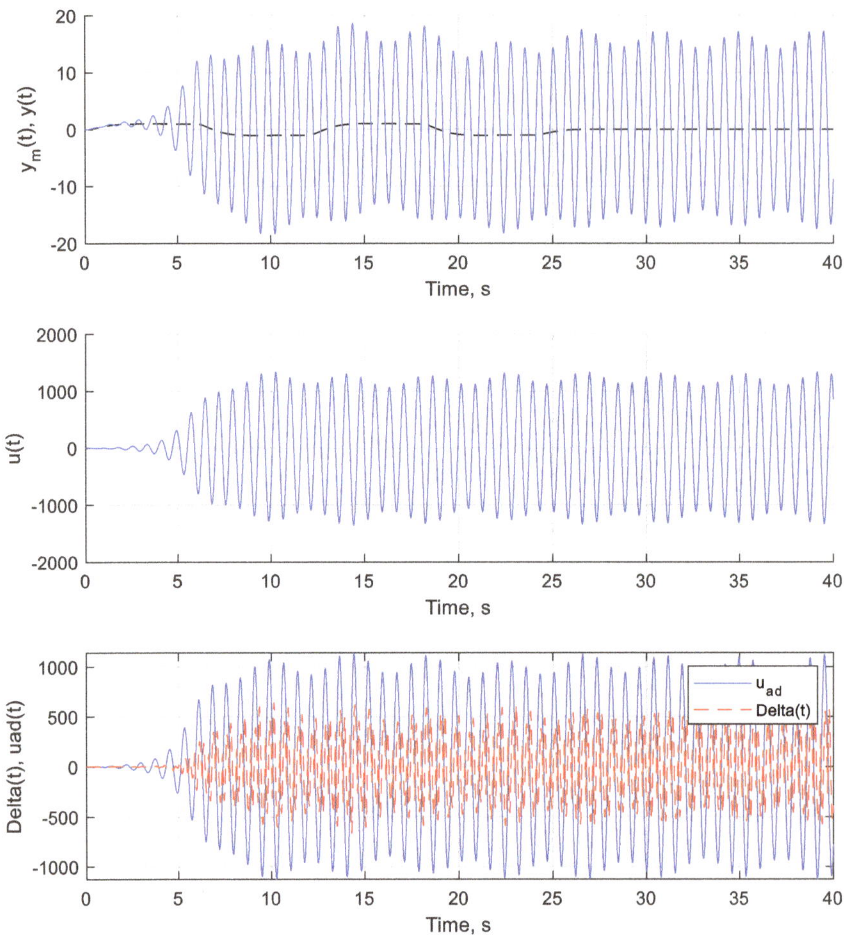

Fig. 5.7 DF-MRAC responses using $\kappa_2 = 0.95\overline{\kappa}_2$.

5.8 DF-MRAC WITH ALR

Adaptive loop transfer recovery (ALR) was previously mentioned in Chapter 3, Remark 3.10 as one of a variety of modification terms than can be employed with DF-MRAC. ALR was originally derived in [26] for use in conventional MRAC applications where the need arises to ensure the loop transfer properties of the reference model are preserved when augmenting the nominal control with an adaptive element in the form of Eq. (5.8). In particular, the goal is to preserve the gain and phase margins of the reference model. In this section, we illustrate the effectiveness of ALR modification by introducing a time delay at the point in the control loop where $u(t)$ is applied to the uncertain system (see Fig. 5.1). The form of the modification term is given in Chapter 3, Remark 3.10, and is easily extended to the output feedback case merely by replacing $x(t)$ with $\hat{x}(t)$ and $e^T PB$ with $\tilde{y}^T(t)$, so the output feedback form of DF-MRAC with ALR becomes

$$\hat{W}(t) = \Omega_1 \hat{W}(t-\tau) + \kappa_2 \left[\beta(\hat{x}(t))\tilde{y}^T(t) - \kappa_m \beta_x(\hat{x}(t))\beta_x^T(\hat{x}(t))\hat{W}(t)\right] \quad (5.69)$$

where $\beta_x(x(t)) \triangleq d\beta(x(t))/dx(t) \in R^{sxn}$. Note Eq. (5.69) can be written in the form of Eq. (3.26) in implementation to avoid the algebraic loop implied by Eq. (5.69). In Section 5.5 we chose $\beta(x(t)) = [1.0,\ \beta_1(x_{1n}(t)),\ \beta_2(x_{2n}(t))]^T$ where $x_{in}(t) = x_i(t)/5$ are normalized inputs and $\beta_i(x_{in}(t)) = (1 - e^{-3x_{in}(t)})/(1 + e^{-3x_{in}(t)})$, $i = 1, 2$ are sigmoidal basis functions. Therefore, in this case,

$$\beta_x(\hat{x}(t)) = 6 \begin{bmatrix} 0 & 0 \\ e^{-3x_{1n}(t)}/(1+e^{-3x_{1n}(t)})^2 & 0 \\ 0 & e^{-3x_{2n}(t)}/(1+e^{-3x_{2n}(t)})^2 \end{bmatrix} /5 \quad (5.70)$$

To best illustrate the effectiveness of DF-MRAC with ALR, we chose $\kappa_2 = \overline{\kappa}/2$ for the design example corresponding to Fig. 5.4a. This example included the

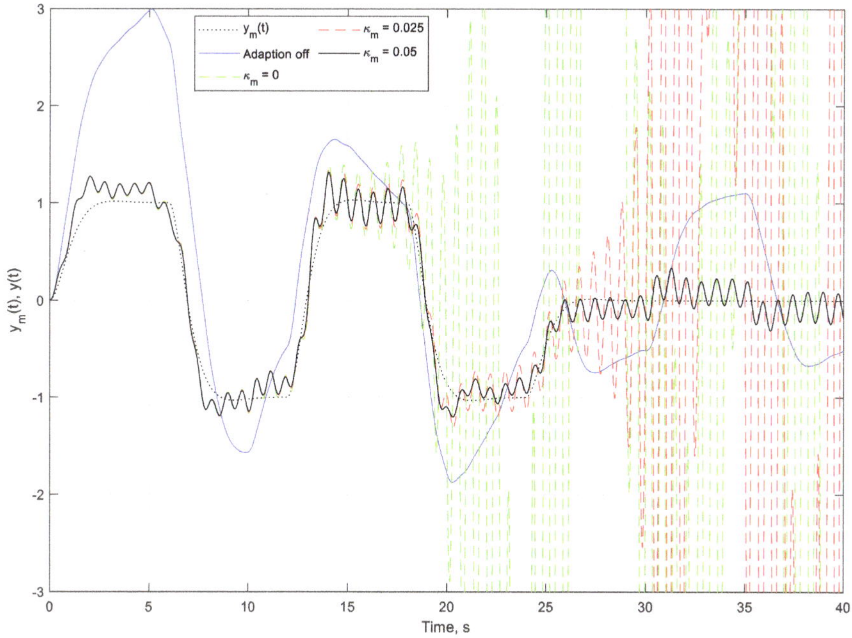

Fig. 5.8 DF-MRAC with ALR applied to the simple example of Section 5.5.

uncertainty and disturbance model in Eq. (5.59), sensor noise, and prefiltering of $\tilde{y}(t)$. We inserted a time delay of 0.25 seconds at both the plant input and the input to the observer in the simulation. Figure 5.8 illustrates the results of four separate runs. The "Adaptation off" response is very similar to the response in Fig. 5.4a obtained without time delay. This shows the nominal controller is highly robust to time delay. The remaining responses show what happens when adaptation is activated with $\kappa_m = 0,\ 0.025,\ 0.05$. The response with $\kappa_m = 0$ (without ALR) is unstable. The response for $\kappa_m = 0.025$ remains stable so long as the command is active (nonzero), after which the response becomes unstable most likely due to a lack of "sufficient excitation." The response for $\kappa_m = 0.05$ remains stable throughout the time interval, and remains so for a large range of values for $\kappa_m > 0.05$.

5.9 SUMMARY

This chapter extends the previously developed state feedback DF-MRAC of Chapter 4 to an output feedback form. As in the full state feedback case, the controller is particularly useful for situations in which the nature of the system uncertainty cannot be adequately represented by a set of basis functions with constant ideal weights, or for situations in which the ideal weights change abruptly, or if there is a difficult-to-model disturbance process. The level of complexity of the output feedback form is lower than many other methods, because it does not require implementation of a reference model when used in a form that augments an existing observer-based linear controller design. Illustrative examples on a simple second-order model show the presented theory, and the simulation results are compatible. In addition, we have presented a simple means by which the effect of sensor noise can be dramatically reduced applicable to all existing forms of adaptive control, and have pointed out the fact that in the simplest case of DF-MRAC, the basis function vector can be chosen as a scalar constant, which results in a linear DF-MRAC law. We also illustrated how ALR modification should be implemented in the output feedback case and illustrated its use in improving robustness of DF-MRAC to the effect of time delay.

REFERENCES

[1] Esfandiari, F., and Khalil, H. K., "Output-Feedback Stabilization of Fully Linearizable Systems," *International Journal of Control*, Vol. 56, No. 5, 1992, pp. 1007–1037.

[2] Krstic, M., Kanellakopoulos, I., and Kokotovic, P., *Nonlinear and Adaptive Control Design*, Wiley, New York, 1995.

[3] Marino, R., and Tomei, P., *Nonlinear Control Design: Geometric,* Adaptive and Robust, Prentice Hall, Englewood Cliffs, NJ, 1995.

[4] Kim, Y., and Lewis, F., *High Level Feedback Control with Neural Networks*, World Scientific, Hackensack, NJ, 1998.

[5] Seshagiri, S., and Khalil, H. K., "Output Feedback Control of Nonlinear Systems Using RBF Neural Networks," *IEEE Transactions on Neural Networks*, Vol. 11, No. 1, 2000, pp. 69–79.

[6] Choi, J., and Farrell, J., "Observer-Based Backstepping Control Using On-line Approximation," IEEE *American Control Conference*, 2000, Chicago, IL, pp. 3646–3650.

[7] Calise, A. J., Hovakimyan, N., and Idan, M., "Adaptive Output Feedback Control of Nonlinear Systems Using Neural Networks," *Automatica*, Vol. 37, No. 8, 2001, pp. 1201–1211.

[8] Hovakimyan, N., Nardi, F., Calise, A., and Kim, N., "Adaptive Output Feedback Control of Uncertain Nonlinear Systems Using Single-Hidden-Layer Neural Networks," *IEEE Transactions on Neural Networks*, Vol. 13, No. 6, 2002, pp. 1420–1431.

[9] Volyanskyy, K., Haddad, W., and Calise, A., "A New Neuro-adaptive Control Architecture for Nonlinear Uncertain Dynamical Systems: Beyond Sigma- and e-Modifications," *IEEE Transactions on Neural Networks*, Vol. 20, No. 11, 2009, pp. 1707–1723.

[10] Ananthakrishnan, S., "Adaptive Tachometer Feedback Augmentation of the Shuttle Remote Manipulator Control System," *IEEE International Conference on Robotics and Automation*, Nagoya, Japan, 1995.

[11] Dash, P., Elangovan, S., and Liew, A., "Design of Nonlinear Expert Supervisory Controllers for Power System Stabilization," Elsevier *Electric Power Systems Research*, Vol. 33, 1995, pp 25–32.

[12] Anderson, C., Hittle, D., Katz, A., and Kretchmar, R., "Synthesis of Reinforcement Learning, Neural Networks and Pi Control Applied to a Simulated Heating Coil," Elsevier *Artificial Intelligence in Engineering*, Vol. 11, 1997, pp. 421–429.

[13] Bodson, M., and Groszkiewicz, J., "Multivariable Adaptive Algorithms for Reconfigurable Flight Control," *IEEE Transactions on Control Systems Technology*, Vol. 5, No. 2, 1997, pp. 217–229.

[14] Wohletz, J., Paduano, J., and Annaswamy, A., "Retrofit Systems for Reconfiguration in Civil Aviation," *AIAA Guidance*, Navigation, and Control Conference, Portland, OR, 1999.

[15] Hoagg, J., Santillo, M. A., and Bernstein, D. S., "Discrete-Time Adaptive Command Following and Disturbance Rejection with Unknown Exogenous Dynamics," *IEEE Transactions on Automatic Control*, Vol. 53, 2008, pp. 912–928.

[16] Lavretsky, E., "Adaptive Output Feedback Design Using Asymptotic Properties of LQG/LTR Controllers," *AIAA Guidance, Navigation, and Control Conference*, Toronto, Canada, 2010.

[17] Kim, K., Calise, A. J., and Yucelen, T., "Parameter-Dependent Riccati Equation-Based Output Feedback Adaptive Control," Wiley *International Journal of Adaptive Control and Signal Processing*, Vol. 31, May 2017, pp. 1608–1622

[18] Wise, K. A., "Applied Controls Research Topics in the Aerospace Industry," *IEEE Conference on Decision and Control*, New Orleans, LA, 1995.

[19] Durham, W., "Constrained Control Allocation," *AIAA Journal of Guidance, Control, and Dynamics*, Vol. 16, No. 4, 1993, pp. 717–725.

[20] Bordignon, K., and Durham, W., "Closed Form Solutions to the Constrained Control Allocation Problem," *AIAA Guidance, Navigation, and Control Conference*, Scottsdale, AZ, 1993.

[21] Haddad, W. M., and Chellaboina, V., *Nonlinear Dynamical Systems and Control. A Lyapunov-Based Approach*. Princeton University Press, Princeton, NJ, 2008.

[22] Potter, J. E., "Matrix Quadratic Solutions," *SIAM Journal on Applied Mathematics*, Vol. 14, 1966, pp. 496–501.

[23] Doyle, J., Glover, K., Khargonekar, P., and Francis, B., "State-Space Solutions to Standard H_2 and H_∞ Problems," *IEEE Transactions on Automatic Control*, Vol. 34, 1989, pp. 831–847.

[24] Singh, S. N., Yim, W., and Wells, W. R., "Direct Adaptive and Neural Control of Wing-Rock Motion of Slender Delta Wings," *Journal of Guidance, Control, and Dynamics*, Vol. 18, 1995, pp. 25–30.

[25] Khalil, H. K., *Nonlinear Systems*, 3rd ed., Prentice Hall, Englewood Cliffs, NJ, 2002.

[26] Calise, A. J., and Yucelen, T., "Adaptive Loop Transfer Recovery," *AIAA Journal of Guidance, Control, and Dynamics*, Vol. 35, No. 3, May–June 2012, pp. 807–815.

CHAPTER 6

Large-Scale Systems

6.1 MOTIVATION

In this chapter we show that the DF-MRAC approach developed in Chapters 3 and 5 can be extended to adaptive control of large-scale interconnected systems, and therefore both matched and unmatched time-varying uncertainties and interconnections can be handled in a decentralized adaptive control setting. A key element in this pursuit is the underlying assumption of DF-MRAC that the weights used to parameterize uncertainty can be time varying. This generalization permits treatment of interconnected systems that may undergo sudden change in each subsystem's uncertain dynamics, such as might be due to damage or assembly and deployment of large, complex orbiting space structures.

6.2 PROBLEM FORMULATION

We consider an uncertain structural system $\mathcal{G}$ consisting of N interconnected subsystems $\mathcal{G}_i$, $i = 1, 2, \ldots, N$. Each subsystem $\mathcal{G}_i$ is described by

$$\dot{x}_i = A_i x_i(t) + B_i[u_i(t) + \Delta_i(t, x_i(t))] + \delta_i(t, x(t)) \tag{6.1}$$

where $x_i(t) \in R^{n_i}$ is the state of $\mathcal{G}_i$, $u_i(t) \in R^{m_i}$ is the control input applied to $\mathcal{G}_i$, $A_i \in R^{n_i x n_i}$ and $B_i \in R^{n_i x m_i}$ are known matrices, and the pair (A_i, B_i) is stabilizable. In addition, $\Delta_i : R_+ x R^{n_i} \to R^{m_i}$ represents matched subsystem bounded disturbances and possibly time-varying matched uncertainty that is subsystem state dependent, and $\delta_i : R_+ x R^{n_1+n_2+\ldots+n_N} \to R^{n_i}$ represents possibly time-varying, total system state dependent subsystem uncertainty and interactions with the other subsystems, where

$$x(t) \triangleq \left[x_1^T(t), x_2^T(t), \ldots, x_N^T(t)\right]^T \tag{6.2}$$

Notice $\delta_i(t, x(t))$ is allowed to be *unmatched*. We assume the controller design for $\mathcal{G}_i$ only has access to $x_i(t)$, and $u_i(t)$ is restricted to the class of admissible controls consisting of measurable functions. Hence, we consider the strictly decentralized control problem.

Assumption 6.1: The matched uncertainty in Eq. (6.1) can be linearly parameterized as

$$\Delta_i(t, x_i(t)) = W_i^T(t)\beta_i(x_i) \quad x_i \in D_{x_i} \tag{6.3}$$

where $W_i(t) \in R^{s_i x m_i}$ is an unknown time-varying ideal weight matrix satisfying $\|W_i(t)\| \leq \overline{w}_i$, $\beta_i : R^{n_i} \to R^{s_i}$. is a basis vector of known functions of the form $\beta_i(x_i) = [b_i, \beta_{1i}(x_i), \beta_{2i}(x_i), \ldots, \beta_{(s_i-1)i}(x_i)]^T \in R^{s_i}$ with a bias component $b_i > 0$, and D_{x_i} is a sufficiently large compact set $D_{x_i} \in R^{s_i}$.

Remark 6.1: As in earlier chapters, due to the fact that time variation is allowed in W_i, Assumption 6.1 encompasses broader classes of uncertainties than those parameterized by constant ideal weights.

Remark 6.2: In Assumption 6.1 we introduced a bias component $b_i > 0$ in $\beta_i(x_i)$. This captures the effect of possibly time-varying external matched disturbances acting on subsystem $\mathcal{G}_i$ that are not state dependent.

Remark 6.3: Although not presented here, the formulation can be extended to the case where there also is uncertainty in B_i, as was done in Section 3.4.

Assumption 6.2: The function $\delta_i(t, x(t))$ in Eq. (6.1) satisfies

$$|\delta_i(t, x(t))| \leq \alpha_i \sum_{j=1}^{N} |x_j(t)|, \quad \alpha_i > 0 \tag{6.4}$$

Remark 6.4: Assumption 6.2 implies the system interconnections satisfy a linear growth inequality, and is standard in the decentralized adaptive control literature. (See, for example, [1] in the case of output feedback and [2] in the case of state feedback.)

6.3 ADAPTIVE CONTROL ARCHITECTURE

Let the feedback control law for subsystem $\mathcal{G}_i$ be

$$u_i(t) = u_{n_i}(t) - u_{ad_i}(t) \tag{6.5}$$

where $u_{n_i}(t)$ is a nominal feedback control given by

$$u_{n_i}(t) = -K_{1_i}x_i(t) + K_{2_i}r_i(t) \tag{6.6}$$

$K_{1_i} \in R^{m_i x n_i}$ is the nominal control gain ensuring

$$A_{m_i} \triangleq A_i - B_i K_{1_i} \tag{6.7}$$

is Hurwitz, and $u_{ad_i}(t)$ is the adaptive feedback control given by

$$u_{ad_i}(t) = \hat{W}_i^T(t)\beta_i(x_i) \tag{6.8}$$

where $\hat{W}_i(t) \in R^{s_i x m_i}$ is an estimate of $W_i(t)$.

Consider the reference model for subsystem $\mathcal{G}_i$, characterizing the desired closed-loop behavior when $\Delta_i(t, x_i(t)) = 0$ and $\delta_i(t, x(t)) = 0$, given by

$$\dot{x}_{m_i} = A_{m_i} x_{m_i}(t) + B_{m_i} r_i(t), \ \lfloor r_i(t) \rfloor \leq \bar{r}_i \tag{6.9}$$

where

$$B_{m_i} \triangleq B_i K_{2_i} \tag{6.10}$$

$x_{m_i}(t) \in R^{n_i}$, and $r_i(t) \in R^{q_i}$, $q_i \leq m_i$ is a bounded piecewise continuous reference input. Because $r_i(t)$ is bounded, it follows that $x_{m_i}(t)$ is upper bounded by $\bar{x}_{m_i}(\bar{r}_i, \varepsilon_i)$ for all $x_{m_i}(0) \in \mathcal{B}_{\varepsilon_i}$.

Define the state predictor [3] as

$$\dot{\hat{x}}_i(t) = A_{m_i}\hat{x}_i(t) + B_{m_i} r_i(t) + L_i[x_i(t) - \hat{x}_i(t)] \tag{6.11}$$

where $L_i \in R^{n_i x n_i}$ ensures

$$A_{e_i} \triangleq A_{m_i} - L_i \tag{6.12}$$

is Hurwitz. Note the state predictor can also serve as a reference model; that is, its dynamics are approximately the same as the reference model in Eq. (6.9) when $\tilde{x}_i \triangleq x_i(t) - \hat{x}_i(t)$ is sufficiently small.

Next, consider the DF-MRAC law for the subsystem $\mathcal{G}_i$ given by

$$\hat{W}_i(t) = \Omega_{1_i}\hat{W}_i(t - \tau_i) + \hat{\Omega}_{2_i}(t) \tag{6.13}$$

where $\tau_i > 0$, and $\Omega_{1_i} \in R^{s_i x s_i}$ and $\hat{\Omega}_{2_i} : R^{s_i} x R^{n_i} \to R^{s_i x m_i}$ satisfy

$$0 \leq \Omega_{1_i}^T \Omega_{1_i} < \kappa_{1_i} * I_{s_i}, \ \ 0 < \kappa_{1_i} < 1 \tag{6.14}$$

$$\hat{\Omega}_{2_i}(t) = \kappa_{2_i}\beta_i(x_i(t))\tilde{x}_i^T P_i B_i, \ \kappa_{2_i} > 0 \tag{6.15}$$

and $P_i \in R^{n_i x n_i}$ is the unique positive definite solution of the algebraic Riccati equation

$$0 = A_{e_i}^T P_i + P_i A_{e_i} - b_i^2 \kappa_{2_i} P_i B_i B_i^T P_i + Q_i \tag{6.16}$$

for any symmetric matrix $Q_i > 0$.

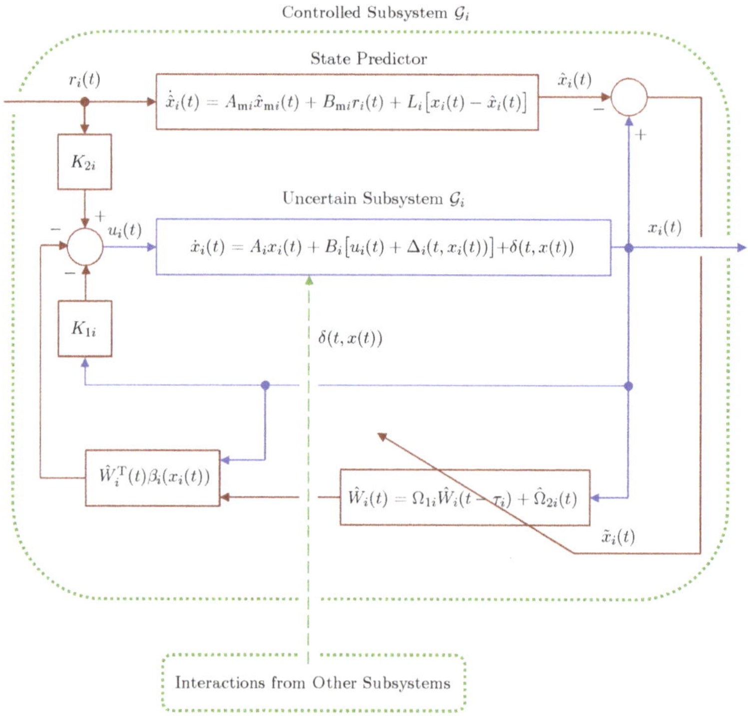

Fig. 6.1 Decentralized DF-MRAC architecture for subsystem $\mathcal{G}_i$.

Remark 6.5: $\tilde{x}_i(t)$ is used in Eq. (6.15) instead of $e_i(t) \triangleq x_i(t) - x_{m_i}(t)$ to estimate $W_i(t)$. If $L_i = 0$, Eq. (6.11) reduces to Eq. (6.9); hence, $\tilde{x}_i(t)$ becomes $e_i(t)$. However, using $\tilde{x}_i(t)$ in Eq. (6.15) instead of $e_i(t)$ adds a low-pass filtering effect to the weight update law. Choosing the state predictor dynamics faster than the reference model dynamics (choosing the eigenvalues of A_{e_i} larger than the eigenvalues of A_{m_i}) aids in suppressing undesired transient behavior [3] and reduces the effect of sensor noise.

Remark 6.6: As noted in Chapter 3, Remark 3.8, DF-MRAC does not employ an integrator in its weight update law. This is advantageous from the perspective of augmenting a nominal controller employing integral action to ensure the regulated output variables track $r_i(t)$ for constant disturbances, regardless of how these disturbances may enter the system.

A visualization of the decentralized DF-MRAC architecture is given in Fig. 6.1.

6.4 STABILITY ANALYSIS

This section presents the main theorem concerning the decentralized DF-MRAC architecture described in Section 6.3. Sufficient conditions needed in its proof involve both the Riccati equation from Eq. (6.16) and the following parameter-dependent Riccati equation associated with the ith subsystem:

$$0 = A_{m_i}^T P_{o_i} + P_{o_i} A_{m_i} + \overline{Q}_{o_i} \tag{6.17}$$

$$\overline{Q}_{o_i} = Q_{o_i} + \mu_i P_{o_i} L_i L_i^T P_{o_i} \tag{6.18}$$

in which $Q_{o_i} > 0$ and $0 < \mu_i < \overline{\mu_i}$, where $\overline{\mu_i}$ is the largest value of μ_i for which there exists a positive definite solution for P_{o_i}. The manner in which $\overline{\mu_i}$ is determined is the same as that explained in Chapter 5, Section 5.3.

Assumption 6.3: $d_{1_i} \triangleq \underline{\lambda}(Q_i) - 3N\alpha_i\overline{\lambda}(P_i) - \ell_i/\mu_i - \sum_{j=1}^N \alpha_j\overline{\lambda}(P_j) > 0$ and $d_{2_i} \triangleq \ell_i\underline{\lambda}(Q_{0_i}) - \sum_{j=1}^N \alpha_j\overline{\lambda}(P_j) > 0$, for an arbitrarily chosen value of $\ell_i > 0$, and by a suitable selection of DF-MRAC design parameters.

Remark 6.7: The conditions in Assumption 6.3 can be achieved by ensuring that $\overline{\lambda}(P_i)$ is sufficiently small. One way this can be done is to make the state predictor dynamics in Eq. (6.11) fast by choosing L_i so that the eigenvalues of A_{e_i} are sufficiently far from the origin in the left half of the complex plane. Similar conditions are required to be satisfied in the decentralized adaptive control literature (see, for example, [4]).

Theorem 6.1: Consider the uncertain system $\mathcal{G}$ consisting of N interconnected subsystems $\mathcal{G}_i$ described by Eq. (6.1) subject to Assumptions 6.1, 6.2, and 6.3. Consider, in addition, the subsystem control laws given by Eq. (6.5), with the nominal controllers given by Eq. (6.6), and with the adaptive controllers given by Eq. (6.8) along with Eqs. (6.11) and (6.13) subject to the conditions in Eqs. (6.14) and (6.15). Then $\tilde{x}_i(t)$, $\tilde{e}_i(t) \triangleq \hat{x}_i(t) - x_{m_i}(t)$, and $\tilde{W}_i(t) \triangleq W_i(t) - \hat{W}_i(t)$ are uniformly ultimately bounded for all $i = 1, 2, \ldots, N$.

Proof: See Section 4 in [5].

6.5 AN EXAMPLE OF THREE INTERCONNECTED SUBSYSTEMS

Consider three inverted pendulums mounted on carts, as depicted in Fig. 6.2. The carts are connected by springs and dampers. Each isolated subsystem is unstable and nonminimum phase, which makes the problem considerably more challenging. The equations of motion of this uncertain system are given by

$$(M + m)\ddot{p}_i(t) + m\ell_p\ddot{\theta}_i(t)\cos(\theta_i(t)) - \mathrm{m}\ell_p\dot{\theta}_i^2(t)\sin(\theta_i(t)) = u_i(t) + \delta_i(t) \tag{6.19}$$

$$m\ell_p\cos(\theta_i(t))\ddot{p}_i(t) + m\ell_p^2\ddot{\theta}_i(t) - mg\ell_p\sin(\theta_i(t)) = 0 \tag{6.20}$$

where M is the mass of each cart (1.0 kg), m is the mass of each rod (0.2 kg), ℓ_p is the distance from the pivot point on each cart to the center of gravity of each rod (0.3 m), g is the gravitational acceleration (9.8 m/s^2), $\delta_i(t)$ is the effect due to

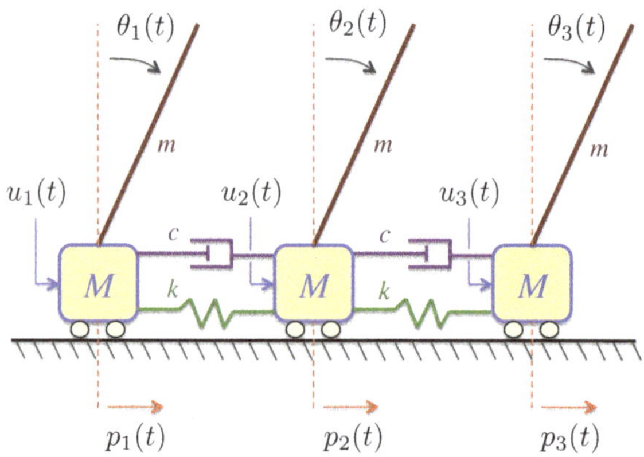

Fig. 6.2 Three inverted pendulums on three interconnected carts.

unmodeled system interconnections represented by

$$\begin{aligned}\delta_1(t) &= k(p_2(t) - p_1(t)) + c(\dot{p}_2(t) - \dot{p}_1(t))\\ \delta_2(t) &= k(p_3(t) - p_2(t)) + c(\dot{p}_3(t) - \dot{p}_2(t))\\ &\quad - k(p_2(t) - p_1(t)) - c(\dot{p}_2(t) - \dot{p}_1(t))\\ \delta_3(t) &= -k(p_3(t) - p_2(t)) - c(\dot{p}_3(t) - \dot{p}_2(t))\end{aligned} \tag{6.21}$$

with $k = 1.4$ N/m, being the spring constant, and $c = 0.7\times10^{-4}$ N-s/m being the damping coefficient. Furthermore, we assume M and m are not known exactly, and are estimated to be $\hat{M} = 2.5$ and $\hat{m} = 0.5$. Finally, we note the system has uncertainties that are matched only for small values of $\theta_i(t)$.

Defining the ith state vector as $x_i(t) = [p_i(t), \dot{p}_i(t), \theta_i(t), \dot{\theta}_i(t)]^T$, linearizing the dynamics in Eqs. (6.19) and (6.20) about the origin, and using $\hat{M}$ and $\hat{m}$ in place of M and m, the matrices A_i and B_i in Eq. (6.1) become

$$A_i = \begin{bmatrix} 0 & 1 & 0 & 0 \\ 0 & 0 & -1.96 & -0.0002 \\ 0 & 0 & 0 & 1 \\ 0 & 0 & 39.2 & 0.0006 \end{bmatrix}, \quad B_i = \begin{bmatrix} 0 \\ 1 \\ 0 \\ -3.33 \end{bmatrix} \tag{6.22}$$

The control objective of each subsystem is to force $p_i(t)$ to track a filtered square wave reference input $r_i(t)$. Each (nominal) subsystem's transfer function from $u_i(t)$ to $p_i(t)$ is unstable and nonminimum phase. The zeros are at –5.71 and 5.71, and the poles are at 0, 0, 6.29, and –6.26.

For this example we chose a P+I form for the nominal controller instead of the form given by Eq. (6.6). For this purpose, we introduce an integrator state $\zeta_i(t)$ and define an augmented state vector $x_{a_i}(t) = [x_i(t) \quad \zeta_i(t)]^T$ for each subsystem whose dynamics are described as

$$\dot{x}_{a_i} = \underbrace{\begin{bmatrix} A_i & 0 \\ [-1\ 0_3] & 0 \end{bmatrix}}_{\overline{A}_i} x_{a_i}(t) + \underbrace{\begin{bmatrix} B_i \\ 0 \end{bmatrix}}_{\overline{B}_i} [u_i(t) + \Delta_i(x_i(t)) + \delta_i(s_i(t))] + \underbrace{\begin{bmatrix} 0_4 \\ 1 \end{bmatrix}}_{B_{m_i}} r_i(t) \tag{6.23}$$

where $s_1(t) = [x_2(t) \quad x_3(t)]^T$, $s_2(t) = [x_1(t) \quad x_3(t)]^T$, and $s_3(t) = [x_1(t)$ $x_2(t)]^T$. In this case, the nominal controller becomes $u_{n_i}(t) = -K_i x_{a_i}(t)$, and $A_{m_i} = \overline{A}_i - \overline{B}_i K_i$. Note the form in Eq. (6.23) is different from the form in Eq. (6.1); however, because $r_i(t)$ is known, it does not affect the stability analysis in [5]. The nominal gain matrix for each subsystem was obtained using linear quadratic regulator theory with $Q = I_5$ and $R = 0.5$, which produced $K_i =$ [-3.79, -4.36, −36.1, -6.12, 1.41]. Figure 6.3 shows the performance of the nominal control system in the absence of parameter uncertainties and interconnections. Figure 6.4 shows the nominal control system is unstable under parameter uncertainties and interconnections.

For each subsystem, the state predictor gain (L_i) in Eq. (6.11) was computed as the optimal Kalman gain corresponding to the stochastic system defined by $\dot{x}(t) = A_{m_i} x(t) + w(t)$ and $y(t) = x(t) + v(t)$, with process noise intensity $100 \times I_5$ and measurement noise intensity $0.2 * I_5$. P_i in Eq. (6.16) was computed with $Q_i = I_5$. The decentralized DF-MRAC was obtained using $\Omega_{1_i} = 0.8$, $\kappa_{2_i} = 2400$, and $\tau_i = 0.2$ seconds. The basis vector was selected as

$$\beta_i(x_i) = \left[1, \tanh(p_i/4), \tanh\left(\dot{p}_i/20\right), \tanh(\theta_i/400), \tanh\left(\dot{\theta}_i/400\right)\right]^T$$

where each state element was normalized by its expected peak value corresponding to the values in Fig. 6.3. For $\ell_i = 0.001$ and $Q_{o_i} = 5000 * I_5$, the conditions in Assumption 6.3 hold when $\alpha_i < 2.53$. In this case, for the forms of the coupling terms in Eq. (6.21), Assumption 6.2 is satisfied as well. Figure 6.5 shows that under parameter uncertainties and interconnections, the decentralized DF-MRAC provides satisfactory performance, comparable to the performance of the nominal system in Fig. 6.3 in the absence of these effects.

Finally consider a case in which the second subsystem is subjected to a time-varying disturbance $d(t) = 0.25 * \sin(0.1t)$ and the spring constant $k(t) = 1.4 + 2.1 * \sin(0.1t)$ is also time varying. Figure 6.6 shows the decentralized DF-MRAC performance for these time-varying uncertainties and disturbances and also depicts $d(t)$. Figure 6.7 shows the corresponding result with $d(t) = 0.25 * \sin(5t)$. Performance is satisfactory in both cases, suggesting the effect of time-varying external disturbances on the regulated outputs is rejected over a broad range of frequencies.

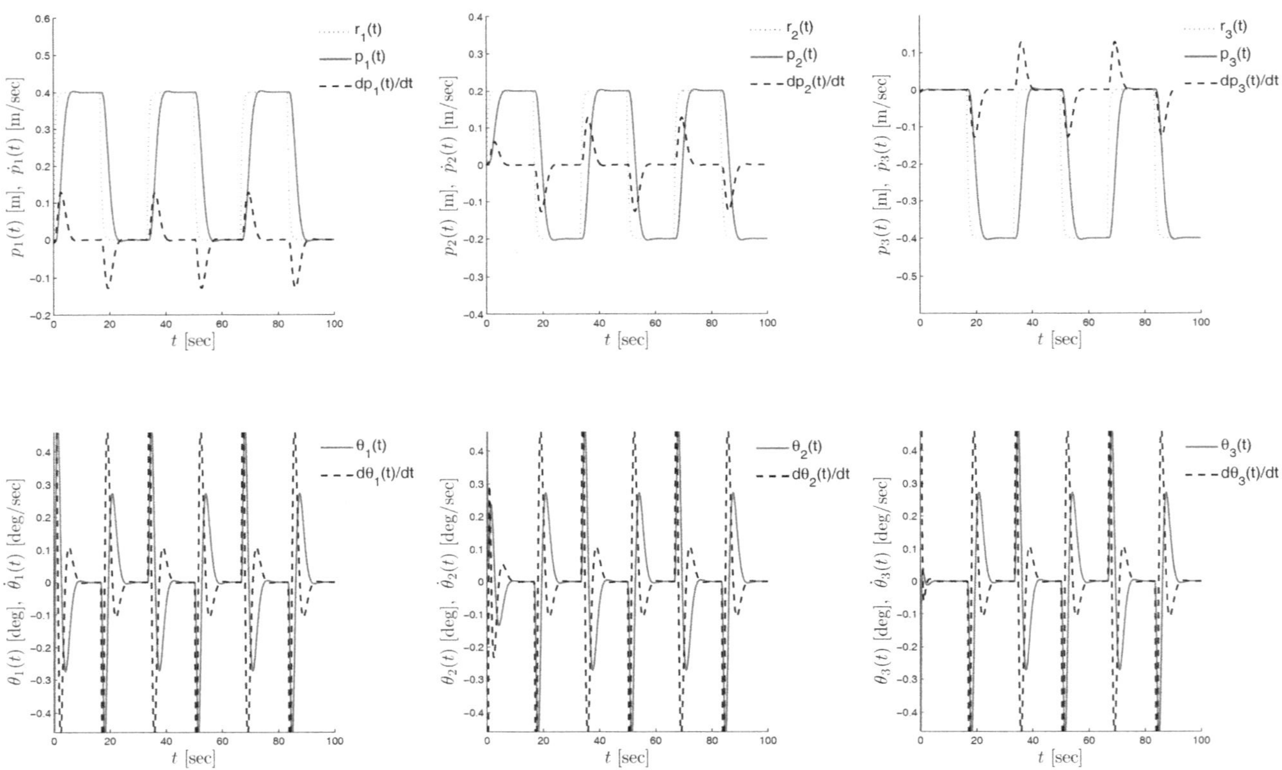

Fig. 6.3 Nominal subsystem responses without parameter uncertainty and interconnections.

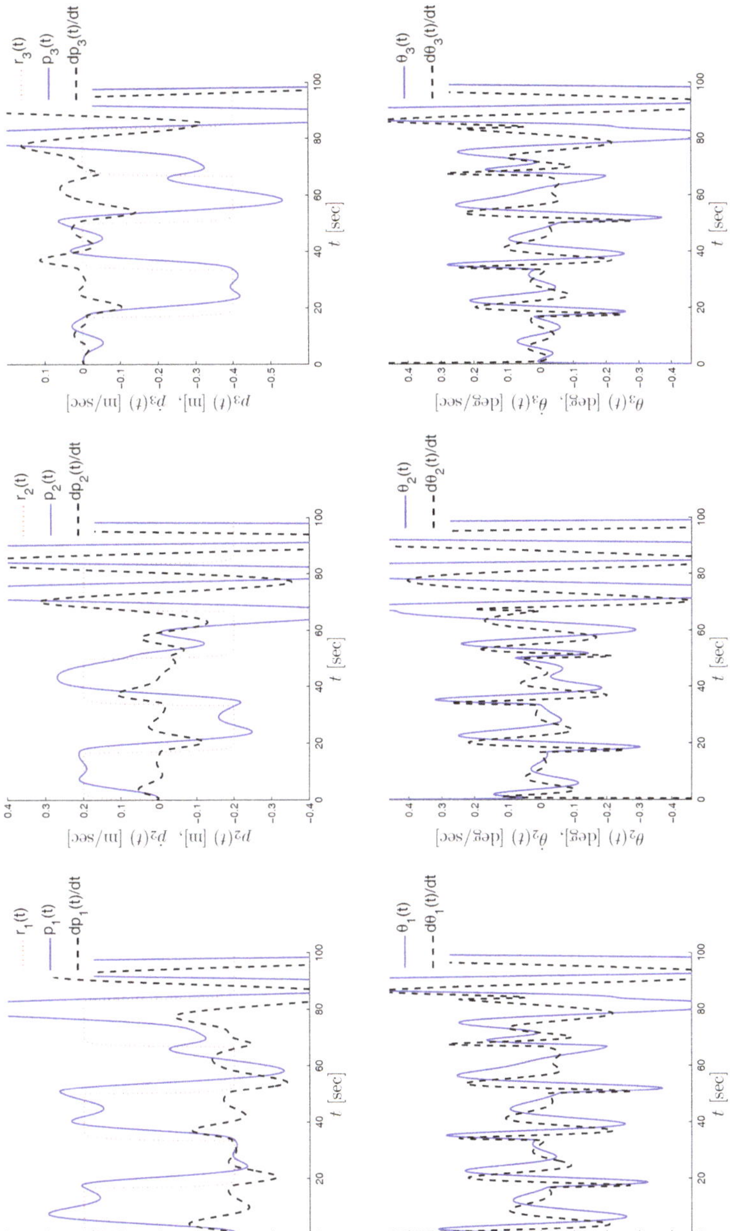

Fig. 6.4 Nominal subsystem responses with parameter uncertainty and interconnections.

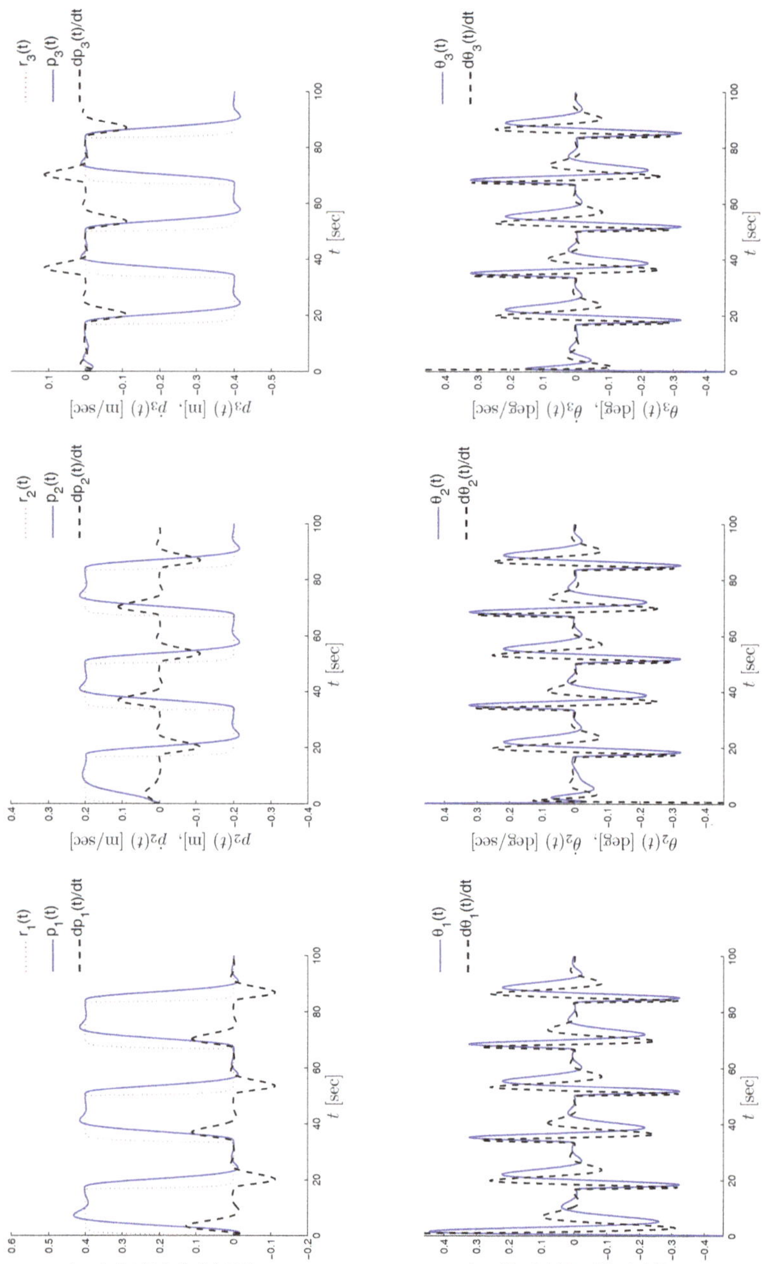

Fig. 6.5 Decentralized DF-MRAC responses with parameter uncertainty and interconnections.

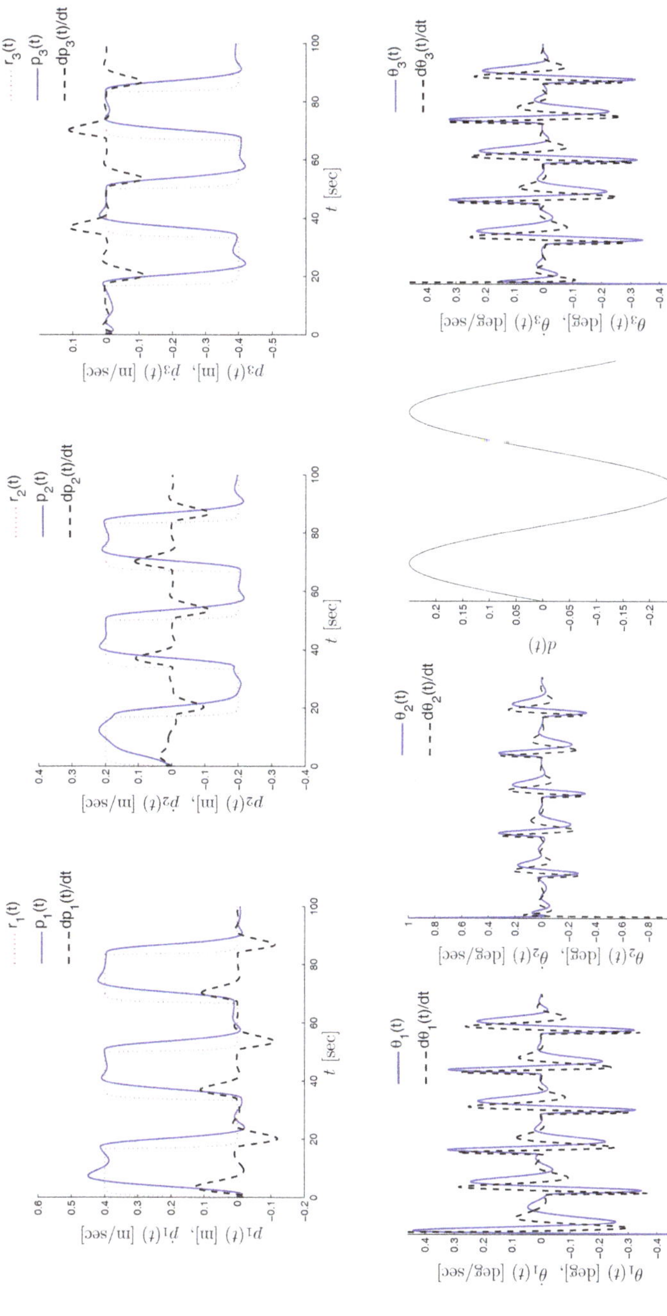

Fig. 6.6 Decentralized DF-MRAC responses with parameter uncertainty, time-varying interconnections, and external disturbance $d_2(t) = 0.25 * \sin(0.1t)$.

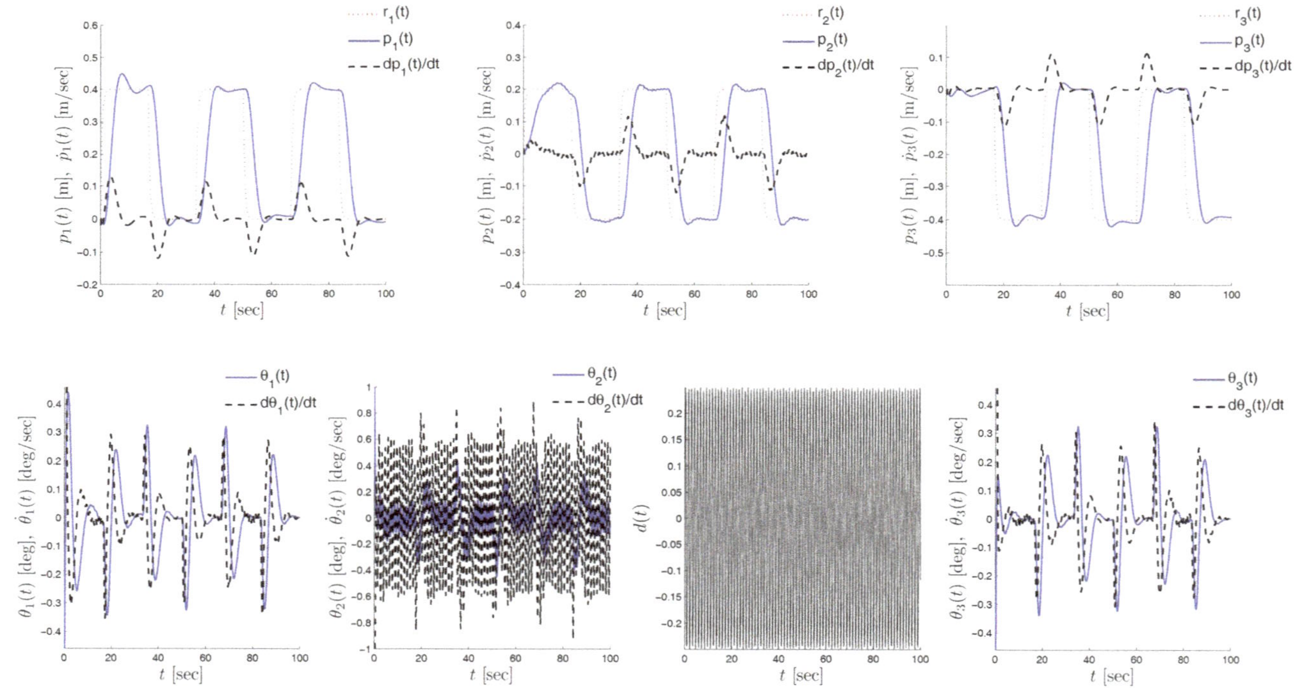

Fig. 6.7 Decentralized DF-MRAC responses with parameter uncertainty, time-varying interconnections, and external disturbance $d_2(t) = 0.25 * \sin(5t)$.

6.6 SUMMARY

This chapter extends DF-MRAC design to a decentralized form for controlling large-scale interconnected systems with both matched and unmatched nonlinear time-varying system uncertainties, interconnections, and matched disturbances. The controller is particularly useful for applications in which uncertain parameters are time-varying or for situations in which external disturbances are difficult to characterize.

REFERENCES

[1] Hovakimyan, N., Lavretsky, E., Yang, B.-J., and Calise, A. J., "Coordinated Decentralized Adaptive Output Feedback Control of Interconnected Systems," *IEEE Transactions on Neural Networks*, Vol. 16, No. 1, 2005, pp. 185–194.

[2] Ioannou, P., "Decentralized Adaptive Control of Interconnected Systems," *IEEE Transactions on Automatic Control*, Vol. 31, No. 4, 1986, pp. 291–298.

[3] Lavretsky, E., "Reference Dynamics Modification in Adaptive Controllers for Improved Transient Performance," *AIAA Guidance, Navigation, and Control Conference*, Portland, OR, 2011.

[4] Gavel, D., and Siljak, D., "Decentralized Adaptive Control: Structural Conditions for Stability," *IEEE Transactions on Automatic Control*, Vol. 34, No. 4, 1989, pp. 413–426.

[5] Yucelen, T., Yang, B.-J., and Calise, A. J., "Derivative-Free Decentralized Adaptive Control of Large-Scale Interconnected Uncertain Systems," Editors: K. Alfriend, M. Akella, J. Hurtado, J. Juang & J. Turner, *Adventures on the Interface of Mechanics and Control*, Tech Science Press, Duluth, GA, pp. 35–60.

CHAPTER 7

Applications to Flight Control

7.1 AEROELASTIC GTM

This section illustrates the application of DF-MRAC in output feedback form, presented in Chapter 5, to an aeroelastic generic transport model (GTM) described in [1]. The model accounts for wing bending and torsion on aircraft performance and stability. The formulation differs mildly from what is presented in Chapter 5 because the nominal controller in this case is defined as a P+I controller. This means if the open loop plant has the form

$$\dot{x}_p(t) = A_p x_p(t) + B_p[u(t) + \Delta(t, x_p(t))] \tag{7.1}$$

$$y_p(t) = C_p x_p(t), \;\; y_{s_p}(t) = C_s x_p(t) + D_s u(t) \tag{7.2}$$

where $y_p(t) \in R^m$ is the regulated plant output variable and $y_{s_p}(t) \in R^{\ell_p}$ is the plant sensed output, then the augmented open loop plant has the form

$$\dot{x}(t) = Ax(t) + B[u(t) + \Delta(t, x_p(t))] + B_r r(t) \tag{7.3}$$

$$y(t) = \overline{C}x(t), \;\; y_s(t) = \overline{C}_s x(t) + \overline{D}_s u(t) \tag{7.4}$$

where $x(t) \triangleq [x(t), \; x_{\text{int}}(t)]^T, \; x_{\text{int}}(t) \in R^m, \; r(t) \in R^m$ is the command vector, and

$$A = \begin{bmatrix} A_p & 0 \\ -C_p & 0 \end{bmatrix}, \; B = \begin{bmatrix} B_p \\ 0 \end{bmatrix}, \; B_r = \begin{bmatrix} 0 \\ I_m \end{bmatrix}, \; \overline{C} = [C_p \; 0],$$
$$\overline{C}_s = \begin{bmatrix} C_s & 0 \\ 0 & I_m \end{bmatrix}, \; \overline{D}_s = \begin{bmatrix} D_s & 0 \\ 0 & 0_m \end{bmatrix} \tag{7.5}$$

For the linearized GTM longitudinal dynamics, $u(t)$ is elevator displacement and $x_p(t) = [\alpha(t), q(t), w(t), \theta(t), \dot{w}(t), \dot{\theta}(t)]^T$, where $\alpha(t)$ denotes angle of attack displacement, $q(t)$ denotes pitch rate, $w(t)$ denotes bending modal amplitude, and $\theta(t)$ denotes torsion modal amplitude. The corresponding matrices associated with the linearized GTM dynamics are:

$$A_p =$$

$$\begin{bmatrix} -0.801 & 0.965 & 1.26\times10^{-2} & 0.509 & 5.46\times10^{-4} & -2.42\times10^{-3} \\ -2.45 & -0.914 & 0.175 & 7.39 & 9.11\times10^{-3} & -3.11\times10^{-2} \\ 0 & 0 & 0 & 0 & 1 & 0 \\ 0 & 0 & 0 & 0 & 0 & 1 \\ 1.42\times10^{3} & 39.6 & -31.3 & -1.40\times10^{3} & -3.25 & -8.26 \\ -2.62\times10^{2} & -5.63 & 3.78 & -1.89\times10^{2} & 0.252 & -3.75 \end{bmatrix} \tag{7.6}$$

$$B_p = [6.50\times10^{-2}\ 3.52\ 0\ 0\ 0\ 0]^T, \quad C_p = [0\ 1\ 0\ 0\ 0\ 0] \tag{7.7}$$

$$C_s = \begin{bmatrix} 0 & 1 & 0 & 0 & 0 & 0 \\ -19.8 & -0.847 & 0.311 & 12.6 & 1.35\times10^{-2} & -6.00\times10^{-2} \end{bmatrix},$$

$$D_s = \frac{U_o}{g}\begin{bmatrix} 0 \\ B_p(1) \end{bmatrix} \tag{7.8}$$

The resulting P+I output feedback DF-MRAC architecture is depicted in Fig. 7.1, which is similar in form to what is shown in Fig. 5.1 in Chapter 5, depicting the case of proportional output feedback DF-MRAC.

The control input and the sensed output are defined as $u(t) = \delta_e(t)$ and $y_s(t) = [q(t),\ A_z(t)]$, where $\delta_e(t)$ is elevator deflection and $A_z(t)$ denotes the normalized acceleration at the aircraft center of gravity. The measurement equation is obtained using the linearized form for $A_z(t)$

$$A_z(t) = U_o[\dot{\alpha}(t) - q(t)]/g \tag{7.9}$$

where U_o is the equilibrium flight speed and g is the acceleration due to gravity. Note from C_p in Eq. (7.7), only $q(t)$ is regulated; therefore, $r(t)$ in Eq. (7.3) is a scalar and $\overline{C}_s \in R^{3\times7}$ in Eq. (7.5). Also, because the second entry of B_p in Eq. (7.7) is approximately 50 times greater than the first, the uncertainty $\Delta(t, x_p(t))$ to be introduced later is approximately matched.

We included the effect of actuator dynamics, measurement noise, and analog filtering of $y_s(t)$ in our simulation. The first-order actuator model for the elevator has a bandwidth of 12 Hz, an amplitude limit of ±30 deg, and a rate limit of ±100 deg/s. To model sensor noise, we assumed independent bandlimited white noise processes with correlation time constants of 0.001 seconds and noise power levels of 1.5×10^{-8} for $q(t)$ and 1×10^{-6} for $A_z(t)$. The first-order

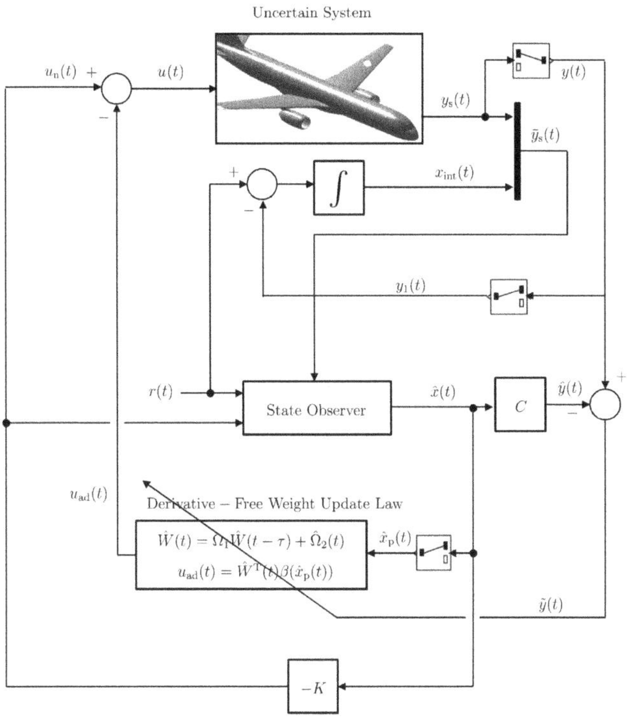

Fig. 7.1 P+I output feedback form DF-MRAC.

sensor filters each have a bandwidth of 40 Hz, well beyond the bandwidth of the nominal control design.

The eigenvalues of the rigid aircraft's short period mode, obtained from the 2×2 upper left matrix partition of Eq. (7.6), are $-0.86 \pm j1.53$. The eigenvalues of the aircraft's two flexible modes, obtained from the 4×4 lower right matrix partition of Eq. (7.6), are $-3.33 \pm j7.57$ and $-0.17 \pm j12.8$. These correspond to a bending and a torsion mode, respectively. The spectrum of A_p is $\rho(A_p) = \{-4.54 \pm j8.07,\ 0.72 \pm j13.1,\ \text{-2.64},\ \text{and}\ 1.56\}$. This shows the linearized model is unstable due to coupling effects and the low damping of the torsion mode.

The frequency responses from the control input to the two sensed outputs for this model are given in Fig. 7.2. The spike in the Bode plot of the transfer function from $\delta_e(t)$ to $q(t)$ is due to a near pole/zero cancellation associated with the torsion mode. The resonance due to the torsion mode is evident in the transfer function from $\delta_e(t)$ to $A_z(t)$.

LQG/LTR theory was used to design the nominal controller [2]. The controller gain matrix K was obtained using $Q = \text{diag}[1,\ 25,\ 0.0001,\ 5,\ 0.0001,\ 5,\ 125]$ to penalize $x(t)$, and $R = 10$ to penalize $u(t)$. The observer gain matrix L was

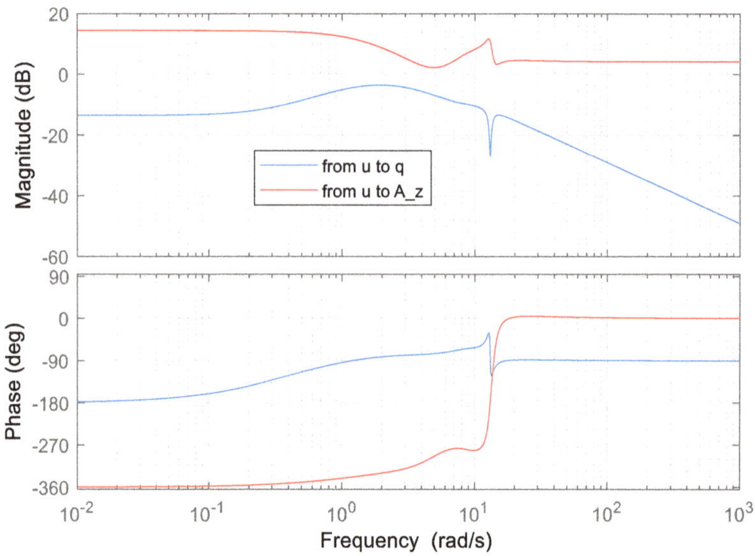

Fig. 7.2 Frequency response of the linearized model.

obtained using $Q_L = \zeta^2 BB^T$ as the process noise matrix, where $\zeta = 6$ is the LTR gain and $R_L = \text{diag}[I_2, 0.01]$ is the measurement noise matrix. The resulting gain matrices for this choice of design parameters are

$$K = [18.6,\ 3.09,\ -0.24,\ 12.5,\ -0.02,\ 0.02,\ -3.53] \tag{7.10}$$

$$L = \begin{bmatrix} 0.45 & 12.5 & 1.45 & 0.21 & 5.42 & -2.25 & -0.23 \\ -0.81 & -16.3 & 1.73 & 0.29 & -42.1 & 10.8 & 0.53 \\ -3.58 & -23.1 & -105 & 0.01 & -168 & 31.9 & 3.53 \end{bmatrix}^T \tag{7.11}$$

Figure 7.3 shows the frequency response of the loop transfer functions with the loop broken at the plant input for both full-state feedback and LQG/LTR loops, where it can be seen the margins of the full-state feedback loop are nearly recovered, although the phase plot differs by approximately 360 deg at low frequencies. For $\zeta \geq 10$, both magnitude and phase curves are essentially identical for all frequencies below 40 rad/s, which is well above the flexible mode frequencies.

Next we consider the command tracking problem with the initial conditions set to zero. Figures 7.4 and 7.5 show the responses with nominal control in the absence of uncertainty, and disturbances. The upper part of Fig. 7.4 compares the pitch rate command $(q_{cmd}(t))$, the reference model response $(q_m(t))$, and the actual plant response $(q(t))$. The lower portion compares the command to the actuator ($\delta_e(t)$ cmd) with the actuator response. The difference cannot be

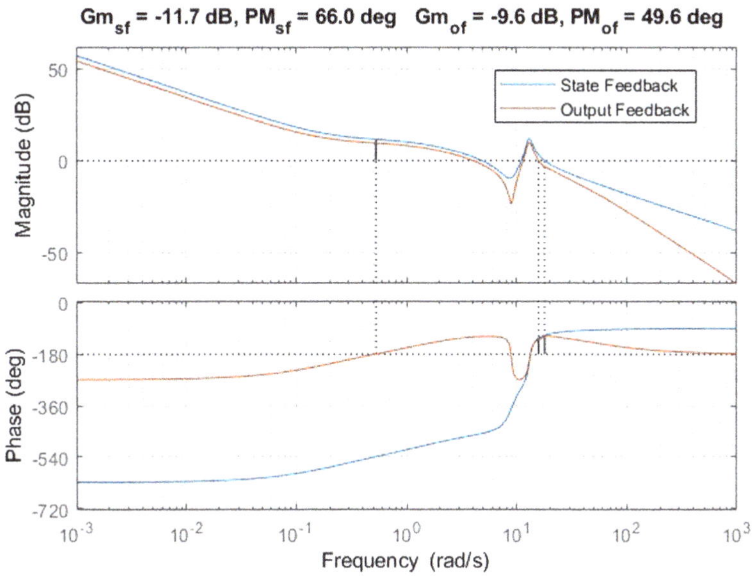

Fig. 7.3 Comparison of frequency responses with loop broken at the plant input.

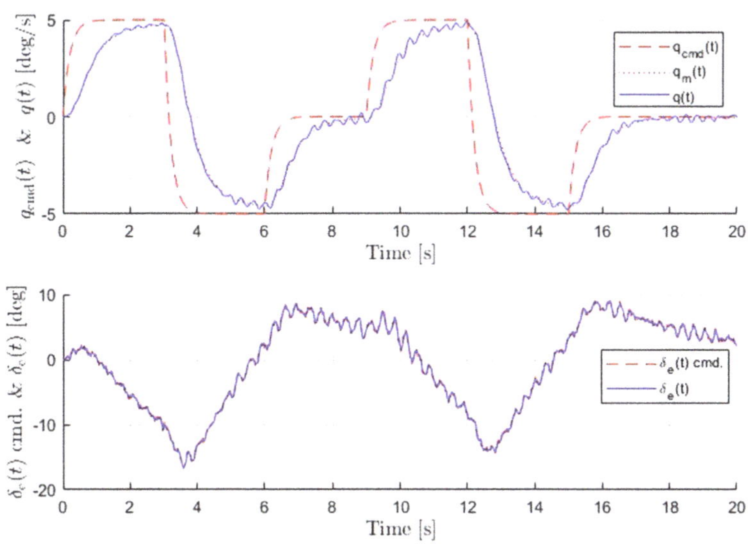

Fig. 7.4 Pitch rate and elevator responses with nominal control without uncertainty.

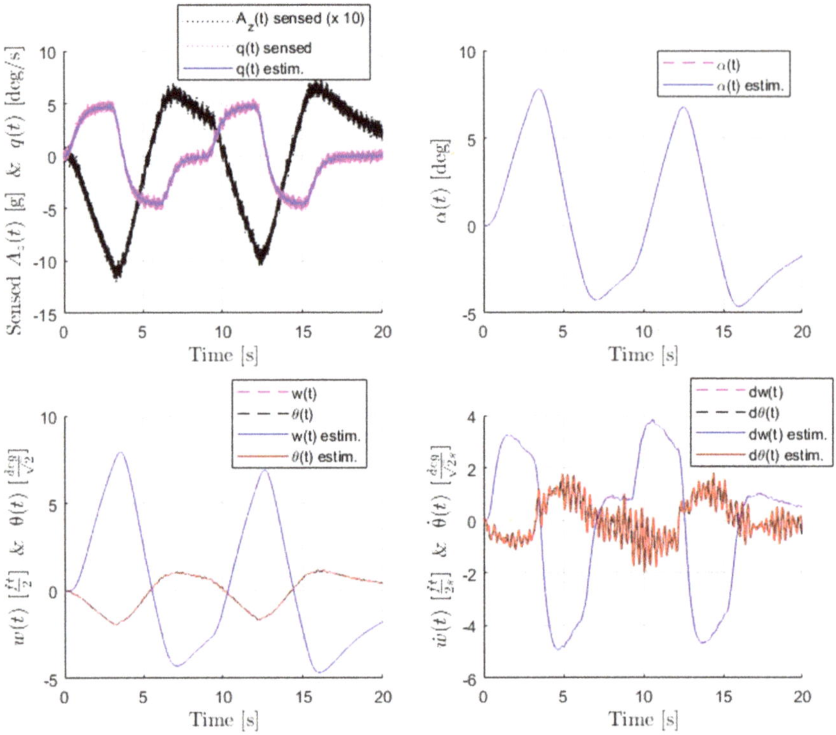

Fig. 7.5 Measurement, state, and state estimate responses with nominal control without uncertainty.

distinguished at this scale. The upper left portion of Fig. 7.5 shows there is a significant amount of sensor noise. The remainder of this figure shows the performance of the observer. The effect of sensor noise is most evident in the rate of change of the torsion amplitude.

For the adaptive control design we used a combination of bias and sigmoidal basis functions of the form

$$\beta(\hat{x}(t)) = [1, \beta_1(\hat{x}_1(t), \quad \ldots, , \beta_6(\hat{x}_6(t))] \tag{7.12}$$

where $\beta_i(\hat{x}_i(t)) = (1 - e^{-\hat{x}_i(t)})/(1 + e^{-\hat{x}_i(t)})$. For this choice of basis functions, $\overline{\beta} = \sqrt{7}$. Selecting $\Omega_1 = 0.5I_7$, the upper bound for μ in Eq. (5.13) is $\overline{\mu} = 3$. Furthermore, for $Q_o = 1.2I_7$, it was determined by Lemma 5.2 in Chapter 5 that $\overline{\nu} = 228.7$, which, in accordance with Theorem 5.1, results in the upper bound $\overline{\kappa}_2 = 32.3$ for κ_2 in Eq. (5.14). For this design we selected $\kappa_2 = 5$. Lastly, we tuned the delay in Eq. (5.12) to $\tau = 0.05$ seconds.

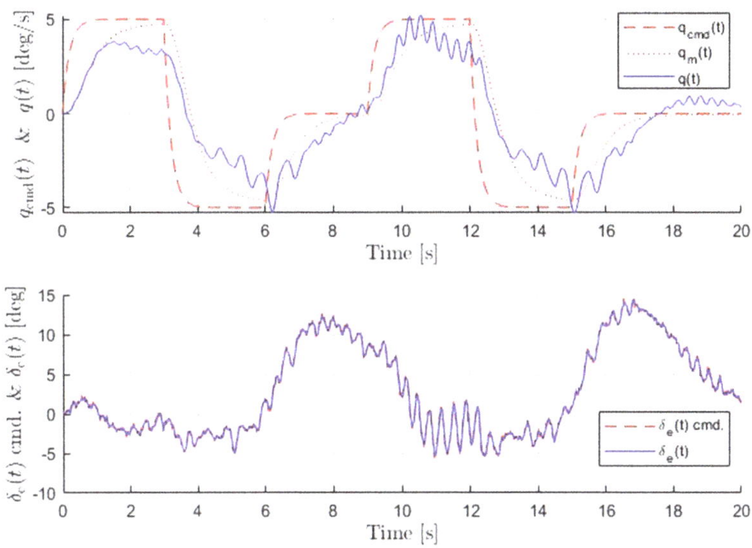

Fig. 7.6 Nominal pitch rate and elevator responses with nonlinear uncertainty.

7.1.1 THE CASE OF NONLINEAR UNCERTAINTY

We first consider the case when there exists a nonlinear uncertainty of the form

$$\Delta(x(t)) = 2.5\alpha(t)q(t) + 2.5\alpha^2(t) + 1.5q^2(t) + 1.25w(t)\theta(t) + 0.25\dot{w}(t)\dot{\theta}(t) + 2.5\theta(t)\dot{\theta}(t) - 2.5\dot{\theta}^2(t) + 0.00025\dot{w}^2(t) \quad (7.13)$$

Figure 7.6 shows the response with nominal control, and Fig. 7.7 shows the DF-MRAC response. Note the improvement in response with DF-MRAC, which is nearly as good as the response observed in Fig. 7.4 without uncertainty. The level of noise present in the elevator command is somewhat greater than the nominal control command shown in Fig. 7.4; however, this can be reduced using the simple filtering method described in Section 5.6 in Chapter 5. This issue will be examined in the next example.

7.1.2 THE CASE OF EXTERNAL DISTURBANCE

Next, a disturbance $d(t) = 0.07\sin(3t)$ is applied to the system through the control input channel. Figure 7.8 shows the response with nominal control, and Fig. 7.9a shows there is a significant improvement in the DF-MRAC response; however, once again we see the level of noise present in the elevator command is significantly increased in the adaptive response. This can be reduced by filtering the error signal used in the adaptive law, as explained in Chapter 5, Section 5.6 and treated in Theorem 5.2. Figure 7.9b shows the DF-MRAC response obtained in

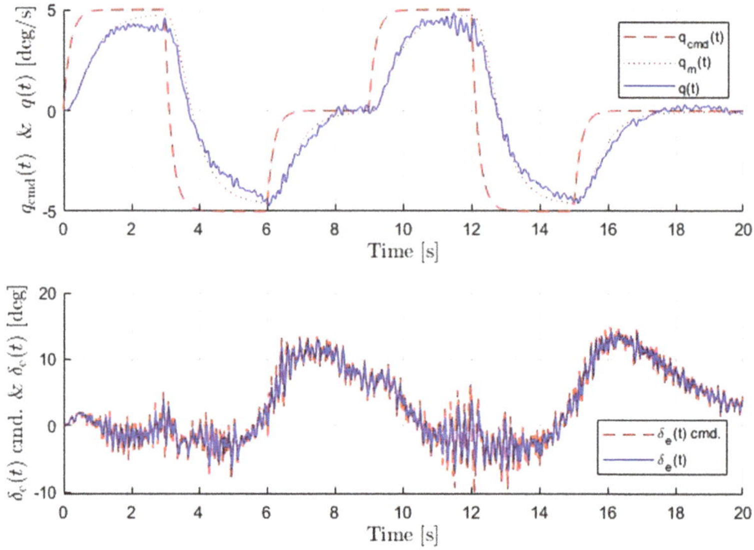

Fig. 7.7 DF-MRAC pitch rate and elevator responses with nonlinear uncertainty.

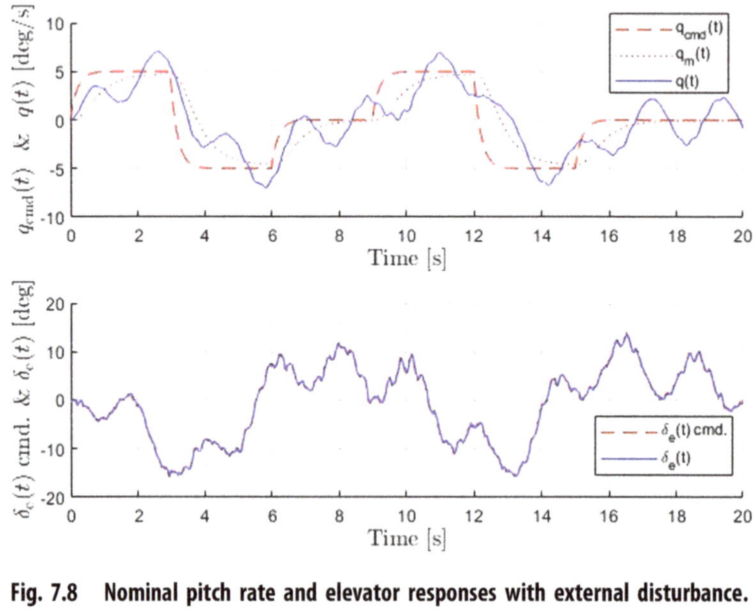

Fig. 7.8 Nominal pitch rate and elevator responses with external disturbance.

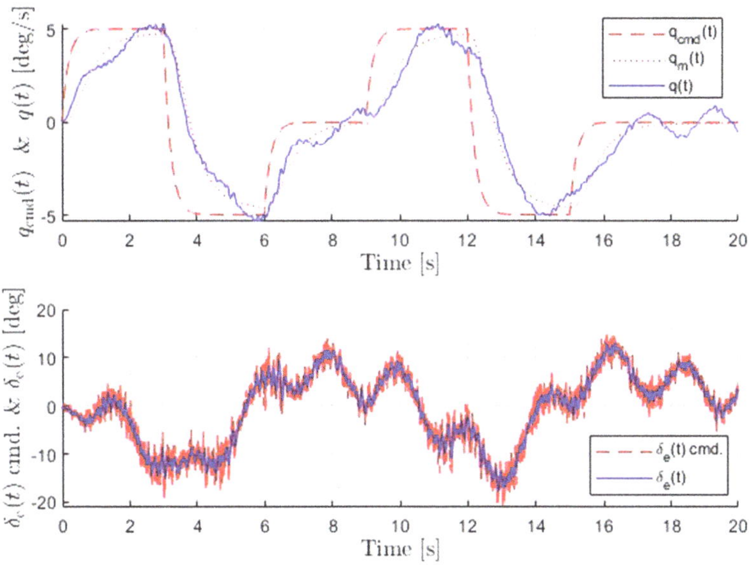

Fig. 7.9a DF-MRAC pitch rate and elevator responses with external disturbance.

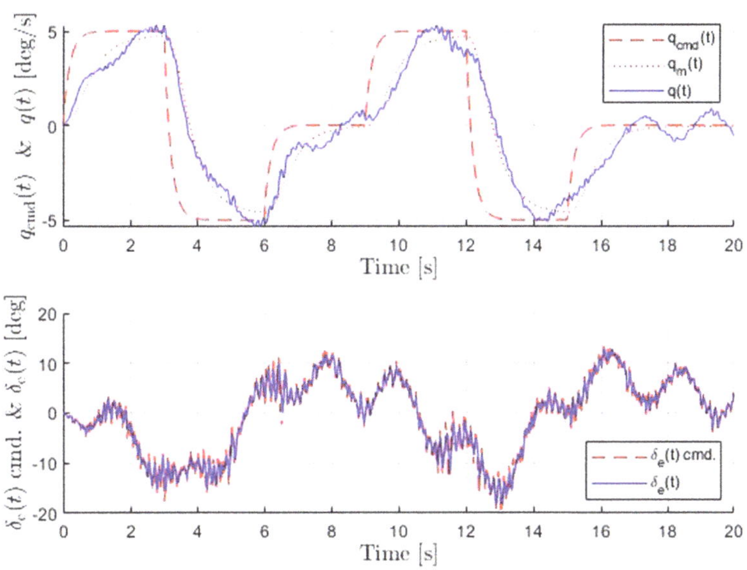

Fig. 7.9b DF-MRAC responses with external disturbance using Eqs. (5.60) and (5.61).

this case when a first-order digital filter is applied in the manner described by Eqs. (5.60) and (5.61), having a filter time constant of 0.005 seconds. All the DF-MRAC results in the remainder of Section 7.1 employ the same digital filter.

7.1.3 THE CASE OF NONLINEAR UNCERTAINTY AND EXTERNAL DISTURBANCE

The uncertainty and external disturbance in the previous two subsections are combined in this next example. The DF-MRAC response is shown in Fig. 7.10. Note the response is very similar to the response shown in Fig. 7.9b.

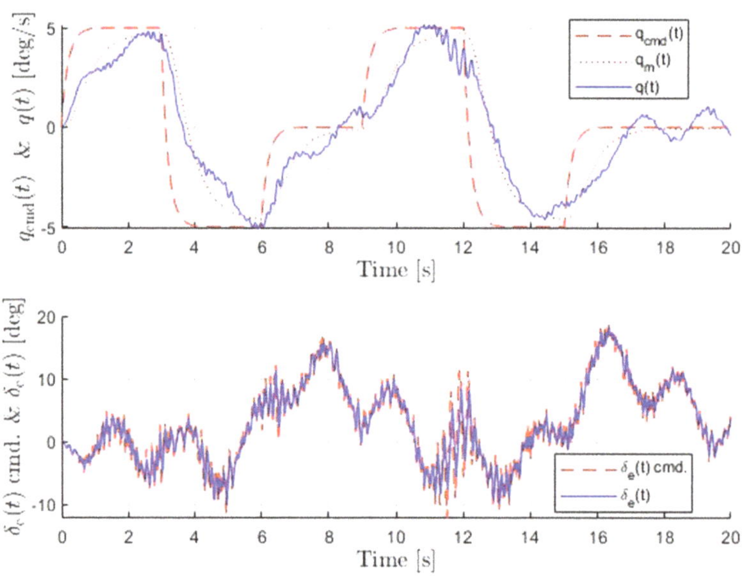

Fig. 7.10 DF-MRAC responses with nonlinear uncertainty and external disturbance.

7.1.4 TWO EXTREME CASES OF PARAMETER UNCERTAINTY

Finally, we illustrate two extreme cases of parameter uncertainty as might be representative of structural damage. In the first extreme case, C_{m_α} changes sign at $t = 6$ seconds. Figure 7.11 shows the nominal control response becomes unstable. Figure 7.12 shows the DF-MRAC response is nearly as good as the case of no uncertainty in Fig. 7.4. In the second extreme case, the entire second row of A_p in Eq. (7.6) changes sign at $t = 6$ seconds. Figure 7.13 shows the nominal control response is nearly unstable. Figure 7.14 shows the effect of this sudden change is hardly noticeable in the DF-MRAC response.

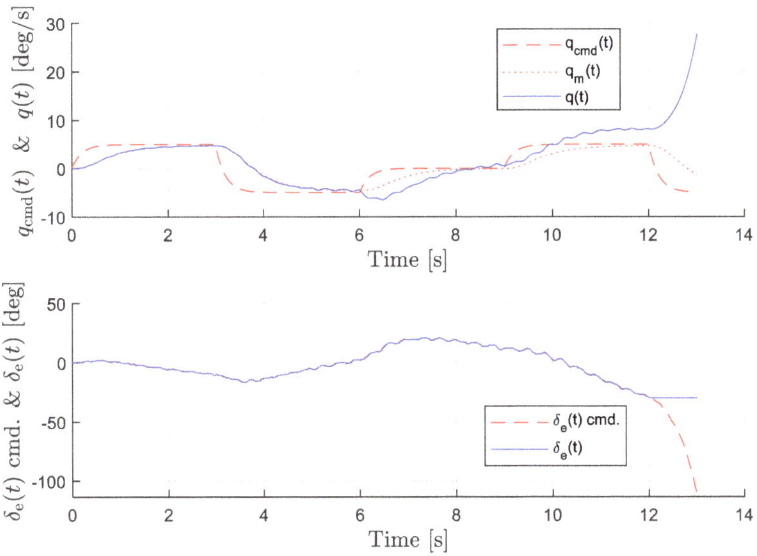

Fig. 7.11 Nominal responses when C_{m_α} changes sign at $t = 6$ seconds.

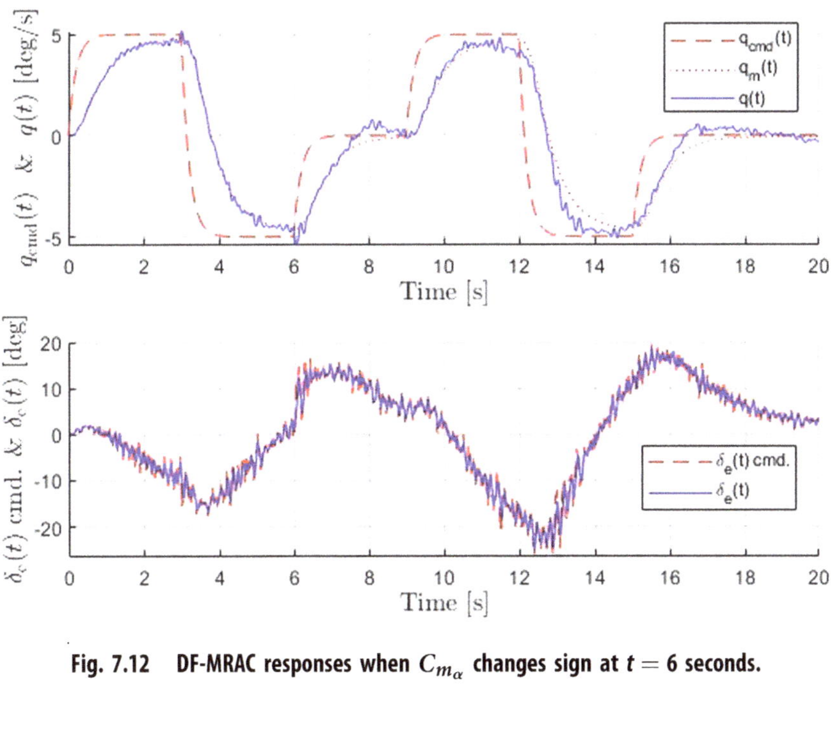

Fig. 7.12 DF-MRAC responses when C_{m_α} changes sign at $t = 6$ seconds.

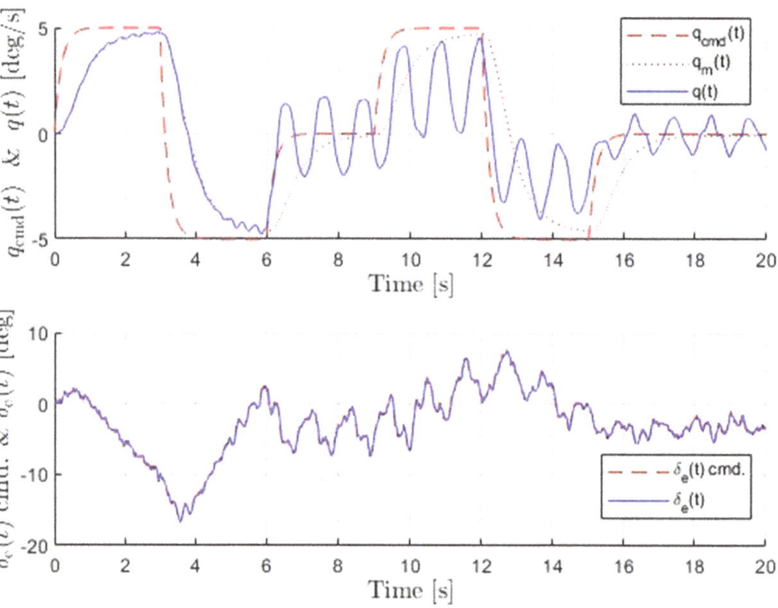

Fig. 7.13 Nominal responses when the second row of A_p changes sign at $t = 6$ seconds.

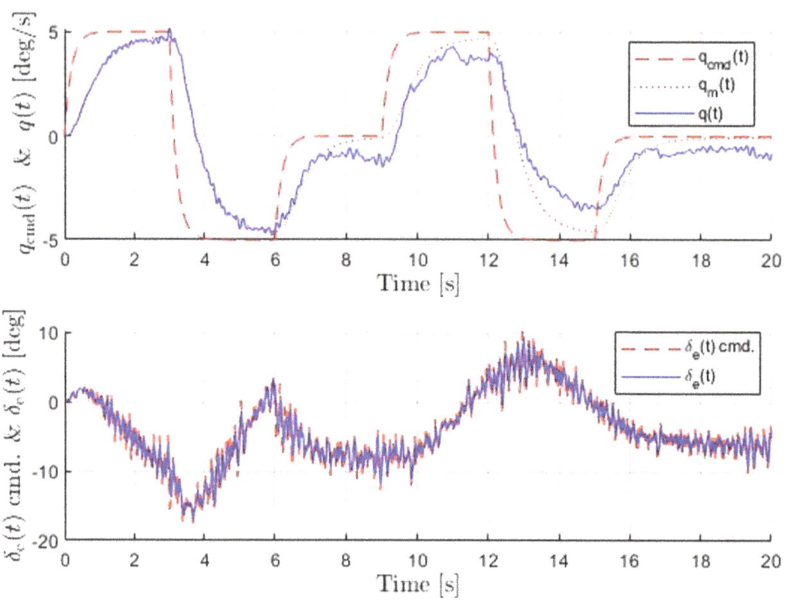

Fig. 7.14 DF-MRAC responses when the second row of A_p changes sign at $t = 6$ seconds.

7.2 TWINSTAR FLIGHT TESTS

Flight testing of adaptive control algorithms was performed at Georgia Tech (GT). This section will summarize the main result of one of these tests focused on adaptive loop-transfer recovery (ALR) to provide robustness of an adaptive system to time delay. The aircraft used in these tests was the GT TwinStar depicted in Fig. 7.15. It carried an embedded autopilot system with an integrated navigation solution used to fuse sensor information by employing an extended Kalman filter. The sensor package included an inertial measurement unit, which provided three-axis angular rate and accelerometer measurements, a global positioning system, and air data sensors [3]. The integrated system provides body attitude, the aircraft's velocity with respect to the air mass, and the flight path angle. TwinStar communicates with a ground station using a 900-MHz wireless data link. The ground station displays onboard information as well as sending commands to the flight control system. Flight measurements of airspeed and throttle setting are also used to estimate thrust. A linear model for the TwinStar in nominal configuration is available in [4].

The form of DF-MRAC with ALR modification was presented in Chapter 5, Section 5.8. A flight test was performed during which a 0.1-second time delay was inserted in the pitch channel of the TwinStar flight control system. During the flight, ALR modification was first on, switched off for a period of approximately 5 seconds, and then switched on again. Figure 7.16 shows the recorded pitch rate during this time period. ALR was switched off at time 2190 and switched on again at 2195. Note the large oscillation in pitch rate during this short time interval, which if allowed to continue would have resulted in an unrecoverable condition.

A video showing the TwinStar behavior during this time period can be viewed on the book's Supporting Materials page on AIAA's Aerospace Research Central (ARC), or at the following link (recommend using Google Chrome): https://drive.google.com/file/d/1a4Xm0FOiUgFZGtn-xUUKe3qwnemvyL4Z.

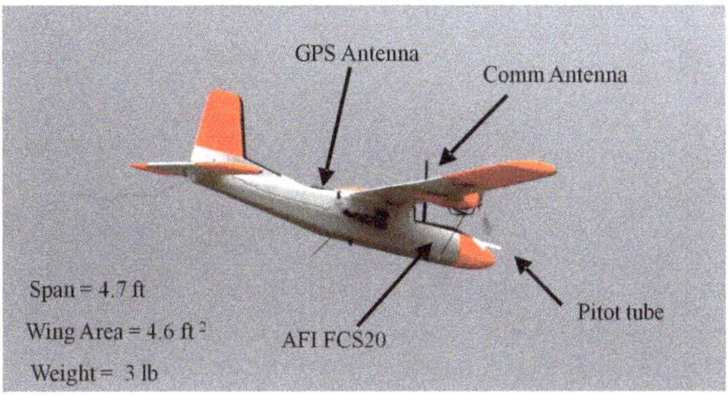

Fig. 7.15 The GT TwinStar aircraft.

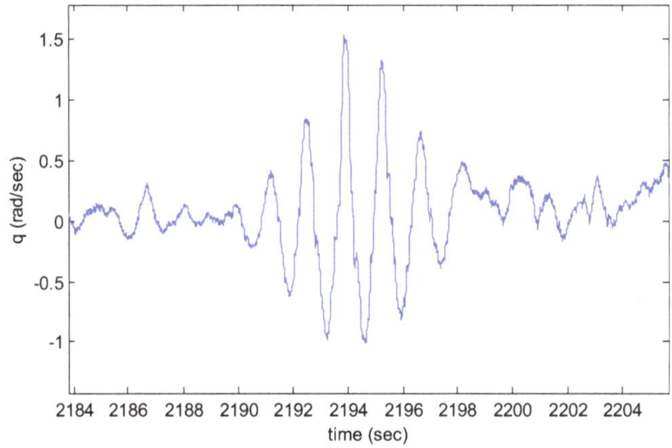

Fig. 7.16 Recorded time history of pitch rate.

7.3 AIRSTAR FLIGHT TEST

In this section we present a limited set of flight test results obtained employing DF-MRAC with ALR modification on the NASA Airborne Subscale Transport Aircraft Research (AirStar) flight test vehicle [5], shown in Fig. 7.17. This form of modification was defined in Remark 3.11 in Chapter 3. Unfortunately, the level of uncertainty introduced in these tests was not sufficient to fully demonstrate the adaptive capabilities of DF-MRAC.

AirStar was designed by NASA to provide a flexible research environment for testing control algorithms under severe uncertainties and failures. In-flight

Fig. 7.17 AirStar flight test vehicle.

damage was represented by introducing additional feedback loops to change stability derivatives representative of unknown destabilizing changes in stiffness and damping terms in the longitudinal and lateral dynamics.

In the following subsections, Figs. 7.18–7.21 present samples of the flight test results obtained. Each figure has four subfigures. The upper two subfigures show roll rate and pitch rate responses for a baseline, nonadaptive control design. The lower subfigures show the roll rate and pitch rate responses for the adaptive design. In all the figures, the dashed line shows the ground pilot commands and the solid line shows the actual AirStar responses.

7.3.1 EMPLOYING BASELINE AND ADAPTIVE CONTROLLERS

Figure 7.18 compares the results when baseline and DF-MRAC controllers are first engaged. Here, the responses of baseline and adaptive controllers in roll and pitch rates look nearly the same because there was no induced uncertainty, and pilot commands occurred primarily in the pitch axis.

7.3.2 ROLL AND PITCH RATE COMMANDS

Figure 7.19 compares the results when a roll rate doublet and a sinusoidal pitch rate are simultaneously commanded, without an induced uncertainty. This case showed improvement in pitch rate response with adaptive control, most likely due to modeling error associated with cross-axis coupling.

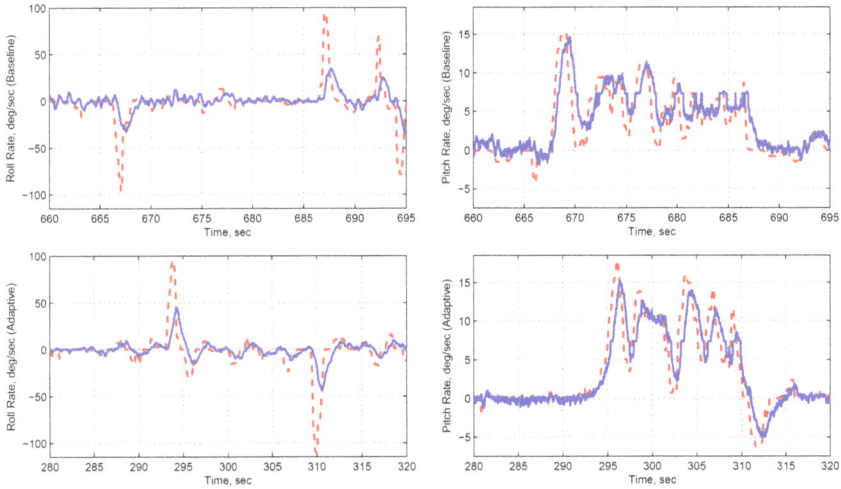

Fig. 7.18 Comparison of baseline and DF-MRAC when first engaged without uncertainty.

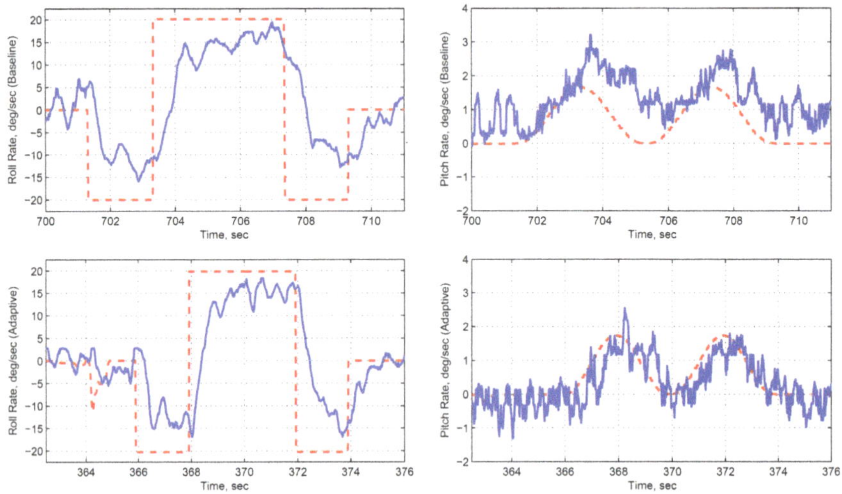

Fig. 7.19 Comparison of responses to roll rate doublet and sinusoidal pitch rate command without uncertainty.

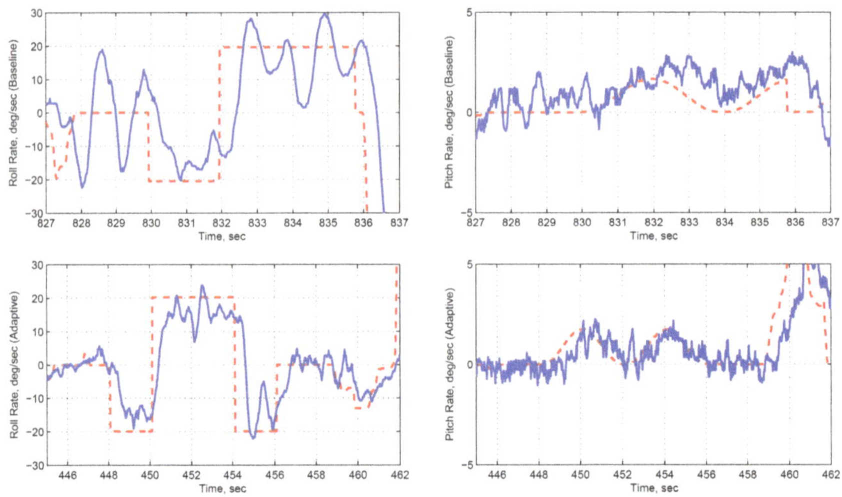

Fig. 7.20 Comparison of responses with uncertainty in C_{m_α} and C_{l_p}.

7.3.3 CHANGES IN C_{m_α} AND C_{l_p}

The effect of changes in C_{m_α} and C_{l_p} under pitch rate command were flight tested. C_{m_α} was reduced in magnitude by 50%, and 0.2 was added to C_{l_p}. Both of these changes have a destabilizing effect in the pitch and roll control channels.

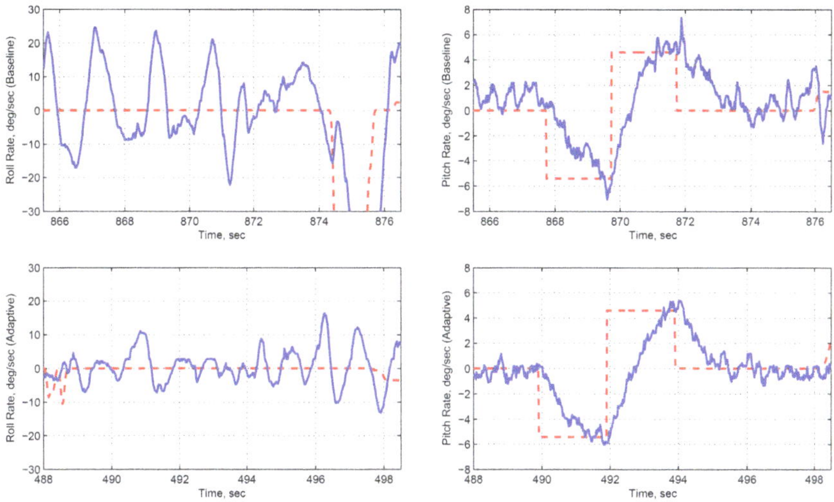

Fig. 7.21 Comparison of responses with increased uncertainty in C_{m_α} and C_{l_p}.

Figure 7.20 compares the AirStar roll and pitch rate responses under these parametric uncertainties for a pitch rate doublet command and a small sinusoidal roll rate command. The uncertainties were artificially induced by feedback of angle of attack and roll rate. The DF-MRAC roll rate response is clearly better than the baseline controller's response. The baseline controller's roll rate response, shown in the upper left side of Fig. 7.20, is much more oscillatory. The DF-MRAC roll rate responses, shown in the bottom left side of Fig. 7.20, show an improvement of more than a factor of 2. The DF-MRAC response in pitch rate exhibited better tracking and was slightly less oscillatory in pitch rate.

Figure 7.21 shows a result for a more severe case of uncertainty in which C_{m_α} was reduced in magnitude by 70%, 0.3 was added to C_{l_p}, and a pitch rate doublet was commanded. In this case, the reduction in pitch rate oscillation is more clearly evident.

REFERENCES

[1] Nguyen, N., Tuzcu, I., Yucelen, T., and Calise, A. J., "Longitudinal Dynamics and Adaptive Control Application for an Aeroelastic Generic Transport Model," *AIAA Guidance, Navigation, and Control Conference*, Portland, Oregon, 2011.

[2] Doyle, J., and Stein, G., "Robustness with Observers," *IEEE Transactions on Automatic Control*, Vol. 24, 1979, pp. 607–611.

[3] Christophersen, H. B., Pickell, W. R., Neidoefer, J. C., Koller, A. A., Kannan, S. K., and Johnson, E. N., "A Compact Guidance, Navigation, and Control System for

Unmanned Aerial Vehicles," *AIAA Journal of Aerospace Computing, Information, and Communication*, 3, May 2006, pp. 187–213.

[4] Debusk, W., Chowdhary, G., and Johnson, E., "Real-Time System Identification of a Small Multi-engine Aircraft," *AIAA Atmospheric and Flight Mechanics Conference*, Chicago, IL, 2009.

[5] Jordan, T. L., and Bailey, R. M., "NASA Langley's AirStar Testbed: A Subscale Flight Test Capability for Flight Dynamics and Control System Experiments," *AIAA Guidance, Navigation, and Control Conference*, Honolulu, Hawaii, 2008.

CHAPTER 8

Conclusions and Some Possible Future Directions

8.1 CONCLUSIONS

The intent of this monograph has been to present the DF-MRAC approach to improve standard designs in adaptive control theory. The key feature is the underlying assumption that the ideal weights are bounded, but otherwise arbitrarily time varying. The approach is particularly useful for situations in which the nature of the system uncertainty cannot be adequately represented by a set of basis functions with unknown constant weights. The system error signals, including the state tracking error and the weight update error, have been shown to be UUB using a Lyapunov-Krasovskii functional without the introduction of modification terms. Furthermore, in Chapter 3 we showed, for the case of state feedback, that the errors approach their ultimate bounds exponentially in time, and we provided expressions defining how the design parameters in the adaptive law can influence the ultimate bound and the exponential rate of convergence to the bound. We have also shown how the derivative-free adaptive law can be modified using many of the existing modification terms, most important of which is ALR modification. All the results have been extended to the case of uncertainty in control effectiveness.

We have shown, through a variety of examples, the superior performance of DF-MRAC in comparison to MRAC when it is evaluated for a variety of situations in which there is a sudden and/or rapidly varying set of dynamics. Moreover, in Chapter 4 we showed that, unlike MRAC, DF-MRAC possesses a natural robustness to unmodeled dynamics, so long as the set of basis functions includes a bias term.

In Chapter 5 we treated the output feedback case, which required the use of a parameter-dependent Riccati equation as an essential ingredient in the stability analysis, and which places a limit on the DF-MRAC adaption gain in order to prove the errors are UUB. In this chapter we also introduced a simple filtering method for avoiding the amplification of sensor noise, which is a common problem in all adaptive control methods. Furthermore we showed it is possible to reduce the basis vector to having just a scalar constant. This greatly simplifies the adaptive law and relieves the designer from having to select a basis vector. This chapter also illustrated the application of ALR in the setting of DF-MRAC on a simple second-order system. Above all, we have shown that the level of complexity of the output feedback form is less than many other approaches in the literature. It allows implementation in a form that augments any observer-based nominal control design, does not require implementation of a reference model, and is applicable to nonminimum phase systems.

Chapter 6 illustrated how DF-MRAC can be applied to large-scale systems, where it is more likely desirable to be able to treat systems that can undergo sudden changes during assembly or that may be subjected to damages after assembly. Allowing the ideal weights to be time varying is particularly advantageous within this setting. Finally, Chapter 7 summarized three applications to aircraft flight control, two of which were flight tested. The results in this chapter confirm the claims regarding DR-MRAC made in earlier chapters on a full-scale problem of practical significance.

8.2 SOME POSSIBLE FUTURE DIRECTIONS

In Chapter 3 we showed, for the state feedback case, that when the ideal weights are constant, the closed loop system errors are UUB to a subspace in which $e(t) = 0$ and $B\tilde{W}^T(t)\beta(x(t)) = 0$ (Theorem 3.2). As a next step, it is desirable to explore the implications of constant ideal weights in the output feedback case. Stability of DF-MRAC with σ- and e-modification terms has been proven only for the case of state feedback. It is desirable to extend Theorem 3.3 to the more general case of output feedback. We recommend the extension in Section 3.4 (Theorem 3.4) to the case of uncertainty at the plant input for the state feedback case to be explored for the output feedback case as well. This extension most likely can be done in a similar fashion.

Although we showed in Chapter 4 that DF-MRAC can be made robust to unmodeled dynamics, the underlying assumption is that the full state of the modeled portion is available for feedback. Therefore, it is of interest to extend the theory of DF-MRAC for systems with unmodeled dynamics to the case in which the modeled portion of the uncertain system has an output feedback form.

Quantification of transient dynamics and gain/time-delay sensitivity at the system inputs and outputs are two open problems in adaptive control theory. To address these problems, there is a need to develop a systematic procedure to assign rates of adaptation and the design matrices used in the Lyapunov or Riccati equations. As discussed, DF-MRAC is shown to have guaranteed transient and steady-state performance bounds; however, there is still a need to quantify the transient dynamics. An analysis based on singular perturbation theory was used to quantify MRAC transients in [1]. A similar analysis can be applied to DF-MRAC, which should provide a similar result.

In order to address gain and phase margins in adaptive systems, we combined DF-MRAC with ALR modification as originally proposed in [2], which is an important step in the direction of certification of adaptive flight control systems. This modification term results in approximate retention of reference model loop properties such as gain, phase, and time-delay margins; however, the use of this modification has only been justified theoretically for the full-state feedback problem. Therefore, a theoretical extension of this approach to all forms of adaptive output feedback control is warranted.

REFERENCES

[1] Lavretsky, E., "Reference Dynamics Modification in Adaptive Controllers for Improved Transient Performance," *AIAA Guidance, Navigation, and Control Conference*, Portland, Oregon, 2011.

[2] Calise, A. J., and Yucelen, T., "Adaptive Loop Transfer Recovery," *AIAA Journal of Guidance, Control, and Dynamics*, Vol. 35, No. 3, May–June 2012, pp. 807–815.

INDEX

SUPPORTING MATERIALS

A complete listing of titles in the Progress in Astronautics and Aeronautics series is available from AIAA's electronic library, Aerospace Research Central (ARC) at arc.aiaa.org. Visit ARC frequently to stay abreast of product changes, corrections, special offers, and new publications.

AIAA is committed to devoting resources to the education of both practicing and future aerospace professionals. In 1996, the AIAA Foundation was founded. Its programs enhance scientific literacy and advance the arts and sciences of aerospace. For more information, please visit www.aiaafoundation.org.

Printed in the USA
CPSIA information can be obtained
at www.ICGtesting.com
BVHW022315200823
668645BV00001B/1

9 781624 107016